AI: APPROACHING INVISIBLE

Surviving and Thriving in a World Where Machines Are Taking Over

NOAH WAYAN

TOP ARTISTS 4U

First Printing, 2025

ISBN

Paperback - 13: 978-1-968871-66-6

Ebook - 13: 978-1-968871-67-3

Top Artists 4 U LLC

15021 Ventura Blvd. #783 Sherman Oaks, CA 91403

TopArtists4U@gmail.com

CONTENTS

To the countless men and women whose voices are fading in the noise of machines - the workers laid off, the families left anxious, the people made to feel invisible in a world racing ahead without them.

This book is for you.

For your courage, your resilience, and your humanity that no algorithm can ever replace. May these pages remind you that even in an age of artificial intelligence, the true measure of progress is how we care for one another.

- Noah Wayan

INTRODUCTION

A New Industrial Revolution

Jasmine sat at her kitchen table, staring at the letter in her hand. The words blurred, not from poor eyesight but from tears she hadn't expected. After twelve years as a customer service representative for a major retailer, she was "no longer needed." The letter was polite, even kind. It congratulated her on her years of service, offered a small severance, and promised "opportunities to reskill." But in plain language, it meant this: a machine now did her job faster, cheaper, and without complaint.She was 52 years old. Not tech-savvy. And terrified. Three states away, Marcus stared at his computer screen in disbelief.

As a freelance copywriter for the past eight years, he'd built a steady client base writing product descriptions, blog posts, and marketing emails. But over the past six months, his clients had quietly begun using AI writing tools. His rates plummeted from $50 per hour to $15, then to nothing at all. The AI could produce his type of content in seconds, not hours. In a suburban classroom outside Phoenix, Sarah wondered if she'd chosen the wrong profession. After fifteen years teaching high school English, she watched as her district introduced AI tutoring systems that could provide instant feedback on essays, answer student questions 24/7, and even create personalized lesson plans. Parents began asking why they needed human teachers when AI could offer unlimited patience and customized learning. Sarah loved her students, but she couldn't shake the feeling that she was becoming obsolete in her own classroom.

These three stories - Jasmine, Marcus, and Sarah - represent millions more playing out across kitchen tables, home offices, and break rooms worldwide. They are the invisible casualties of the greatest economic transformation since the Industrial Revolution. While headlines celebrate AI breakthroughs - self-driving cars, medical diagnoses, artistic creation - there is a quieter, more painful reality unfolding in living rooms everywhere. Jobs are vanishing not in decades, but in months. And for those who cannot quickly adapt, the world feels like it is spinning beyond their reach.

The Promise and the Threat

Artificial Intelligence has been called "the new electricity." Just as electricity transformed every aspect of society a century ago - from how we light our homes to how we power our factories - AI is rapidly infiltrating nearly every industry. The promises are dazzling: personalized medicine that could cure previously untreatable diseases, productivity boosts that could solve global poverty, autonomous systems that could eliminate traffic deaths, and creative tools that could democratize art and writing.

McKinsey & Company estimates that AI could contribute up to $13 trillion to global economic output by 2030. The World Economic Forum predicts AI will create 97 million new jobs by 2025. Tech leaders speak breathlessly of an age where human creativity is unleashed from mundane tasks, where we're free to pursue higher-order thinking while machines handle the routine work.

But here's the darker truth that doesn't make it into the glossy presentations: unlike past technological revolutions, this one isn't just automating physical labor - it's automating thinking, communication, and creativity. It's not just factory jobs at risk anymore, but office jobs, creative jobs, teaching positions, driving careers, and even managerial roles. This means the middle class, once thought safe from automation, suddenly finds itself on unstable ground.

The statistics tell a stark story. Oxford Economics research suggests that 20 million manufacturing jobs worldwide could be lost to automation by 2030. But that's just the beginning. PwC analysis indicates that up to 30% of existing jobs could be at risk of automation by the mid-2030s, with the highest risk in sectors like transportation, manufacturing, and retail. More troubling still, a 2023 Goldman Sachs study found that AI could automate 25% of work tasks across the U.S. and Europe, affecting 300 million full-time jobs. These aren't just numbers on a spreadsheet - they represent human lives, families, and communities.

Why This Time Is Different

Throughout history, technological advancement has displaced workers, but new opportunities have emerged. During the Industrial Revolution, people displaced from agricultural work found employment in factories. When factory automation took hold, workers moved into offices, services, and creative industries. The pattern seemed clear: technology eliminated some jobs but created others, often better ones. This time feels different for four crucial reasons:

1. The Speed of Change Previous technological shifts unfolded over decades, giving workers and institutions time to adapt. The transition from agriculture to manufacturing took nearly a century. The shift to service economies happened over fifty years. AI's impact is compressed into years, not decades. ChatGPT reached 100 million users in just two months - faster than any technology in human history.

The speed at which AI capabilities are advancing leaves little time for retraining, policy adjustment, or social adaptation.

2. The Scope of Impact While past automation primarily affected physical tasks, AI targets cognitive work - the kind that has been the foundation of middle-class prosperity. Legal research, financial analysis, content creation, customer service, and even medical diagnosis are all within AI's reach. This means the disruption isn't limited to blue-collar workers; it extends to lawyers, accountants, journalists, doctors, and teachers.

3. The Concentration of Benefits The wealth generated by AI is flowing to a relatively small number of technology companies and their investors. Unlike the Industrial Revolution, which created new categories of employers and workers, AI's benefits are concentrating among those who own the algorithms, data, and computational infrastructure. This creates a winner-take-all dynamic that could deepen inequality rather than broadly distributing prosperity.

4. The Learning Curve Perhaps most critically, AI can learn and improve at a pace that humans cannot match. While a human worker might take years to master a skill, an AI system can be trained on millions of examples in hours. This means that even as new jobs emerge, AI may master them before human workers have a chance to adapt.

Why This Book Matters

This book is not about glorifying AI or demonizing it. It's not about fighting the inevitable or pretending that technological change can be stopped. Instead, it's about facing reality with honesty, courage, and hope. It's about equipping ordinary people - not tech executives or policy experts, but the forgotten majority - with practical tools to navigate this transformation. We will examine this challenge from every angle, with no sugarcoating:

The jobs and industries most at risk - and which ones offer greater security in an AI-dominated future

Practical adjustments for people who aren't "tech natives" - strategies that don't require becoming a computer programmer

Psychological and emotional tools for managing anxiety, rebuilding identity, and finding purpose in uncertain times

The systemic changes needed from governments, corporations, and communities to ensure this transition doesn't leave millions behind

Preparation strategies for the next generation - so our children can thrive in a world where AI is simply part of the landscape

Most importantly, this book offers a framework for maintaining human dignity and purpose in a world increasingly dominated by machines. Because while AI may be approaching invisibility - becoming so embedded in our daily lives that we barely notice it - the human response to it doesn't have to be invisible. We can face this challenge with intention, community, and hope.

The Choice Before Us

This is the story of humanity at a crossroads. We stand at a moment that will define the next century of human experience. The choices we make in the next decade - as individuals, communities, and societies - will determine whether AI becomes a tool for broad human flourishing or an engine of unprecedented inequality. One path leads to a future where millions are left behind, where prosperity concentrates among a technological elite while the majority struggles with displacement, anxiety, and economic insecurity. This is the path of passive acceptance, where we allow AI to shape us rather than shaping how we use AI.

The other path leads to adaptation, resilience, and a redefinition of human value that goes beyond what machines can measure. This is the path of intentional response, where we harness the benefits of AI while protecting the dignity and purpose of human work and life. The question facing every reader of this book is both simple and urgent: In the Age of AI, will ordinary people be partners in progress or casualties of it? The answer depends on what we do next.

PART I: THE RECKONING

1

THE INVISIBLE CASUALTIES

The First to Fall

Late one evening in Akron, Ohio, James parked his eighteen-wheeler for what he didn't know would be the last time. After nearly thirty years of long-haul trucking, he'd seen a lot of changes - new regulations, electronic logging devices, GPS navigation systems that replaced the paper maps he'd once relied on. But he'd always adapted. Truckers were resourceful by nature, and James took pride in his ability to handle whatever the road threw at him. The warnings about self-driving trucks had been circulating for years, but James, like many in his profession, had dismissed them.

How could a machine handle black ice on a Wyoming interstate at 2 AM? How could it navigate the complex loading docks of urban distribution centers? How could it deal with the hundred small decisions that kept goods moving safely across the country? The answer came gradually, then suddenly. First, his company introduced "platooning" systems where trucks would follow each other in AI-controlled convoys, reducing fuel consumption and increasing safety on highway stretches. James found himself driving alone for shorter segments while the computer handled the long, straight portions of interstate. Then came the terminal-to-terminal autonomous routes, where trucks would drive themselves between major hubs while human drivers handled the "first and last mile" connections.

Within three years, James's role had been reduced to moving trucks around parking lots and handling local deliveries. His pay dropped from $65,000 a year to $32,000. The layoffs came in waves - first the older drivers, then the ones with less-than-perfect safety records, then everyone except a skeleton crew of technicians who could troubleshoot the autonomous systems. At 56, James found himself scrolling through job boards for the first time in decades, competing with workers half his age for warehouse positions that paid $15 an hour and offered no benefits.

The skills that had once made him valuable - knowledge of routes, ability to handle difficult weather, customer relationships built over years of reliable service - were suddenly worthless in a world where algorithms optimized every mile and machines never got tired. James isn't alone. His story echoes in call centers from Manila to Milwaukee, in newsrooms from New York to Nashville, and in offices from Seattle to San Antonio. They are the invisible casualties of the AI revolution - workers who don't make headlines but whose quiet struggles reveal the true cost of progress.

The Expanding Scope of Displacement

The narrative around AI and jobs has evolved rapidly. Early discussions focused on manufacturing and routine manual labor - jobs that seemed obvious targets for automation. But as AI capabilities have advanced, the scope of potential displacement has expanded far beyond what most experts predicted even five years ago.

Customer Service: The Canary in the Coal Mine

Customer service was among the first white-collar sectors to experience large-scale AI displacement. The transformation began with simple chatbots that could answer basic questions, but quickly evolved into sophisticated systems capable of handling complex customer interactions. Take the case of TechSupport Solutions, a company that once employed 1,200 customer service representatives across three call centers. In 2019, they began implementing an AI system that could resolve 70% of customer inquiries without human intervention. By 2022, they had reduced their workforce to 300 employees, mostly supervisors and specialists handling the 30% of cases the AI couldn't resolve.

Maria Rodriguez, who worked for TechSupport for eight years, watched her department shrink from 50 people to 12. "The AI could pull up customer histories, check account statuses, and even process refunds faster than we could," she recalls. "At first, we thought it would just help us be more efficient. We didn't realize it would replace us entirely." The human cost became clear in the performance metrics. While customer satisfaction scores remained stable and response times improved, the displaced workers struggled to find comparable employment. Many took jobs in retail or food service at half their previous wages. Others left the workforce entirely, contributing to what economists call "hidden unemployment" - people who stop looking for work because they believe no suitable jobs are available.

Administrative Work: The Quiet Revolution

Perhaps no sector has experienced more dramatic yet invisible change than administrative work. Executive assistants, data entry clerks, bookkeepers, and office coordinators have seen their roles transformed or eliminated as AI systems take over tasks that once required human judgment and coordination. Consider the evolution of scheduling and calendar management. What once required a skilled administrative assistant who could balance competing priorities, understand unspoken preferences, and navigate complex interpersonal dynamics can now be handled by AI systems that can access multiple calendars, book conference rooms, send invitations, and even reschedule meetings when conflicts arise.

Jennifer Walsh worked as an executive assistant to a team of five marketing directors at a Fortune 500 company. Over her 12 years there, she had developed an intricate understanding of each executive's working style, travel preferences, and personal commitments. She could anticipate needs before they were voiced and solve problems before they became crises.

> "I thought my institutional knowledge and personal relationships made me irreplaceable," Jennifer explains. "But when the company implemented an AI-powered 'executive support system,' they discovered that most of what I did could be automated. The AI could book travel, manage calendars, and even draft emails in each executive's voice. The personal touch I provided was seen as nice to have, but not essential."

Jennifer's position was eliminated along with 60% of the administrative support roles at her company. At 48, she found herself competing for entry-level positions with recent college graduates, despite her decades of experience.

Content Creation: The Creative Disruption

The impact on creative and content-creation fields has been particularly jarring because these were supposed to be "safe" from automation. Writing, design, and multimedia production were thought to require uniquely human creativity and emotional intelligence. That assumption has been challenged by AI systems capable of generating compelling written content, stunning visual designs, and even music compositions. Marcus Chen built a successful freelance copywriting business over eight years, specializing in product descriptions, blog posts, and email marketing campaigns for small and medium-sized businesses. He prided himself on understanding each client's brand voice and creating content that felt authentic and engaging.

The shift began subtly. Clients started asking for faster turnarounds and lower prices, mentioning that they were "exploring other options." Marcus initially assumed they were working with cheaper offshore writers. It wasn't until a long-term client showed him an AI-generated product description that he realized the nature of the competition.

> "The AI version wasn't perfect," Marcus remembers, "but it was 80% as good as what I would have written, and it took five minutes instead of five hours. When you're a small business owner trying to manage costs, that math is pretty compelling."

Marcus's client base eroded gradually over 18 months. His hourly rate dropped from $75 to $25 as he tried to compete on price. Eventually, he was reduced to editing and improving AI-generated content rather than creating original work. The irony wasn't lost on him - he was now training the very systems that were replacing him.

Transportation: The Road to Displacement

The transportation industry, which employs millions of Americans in driving roles ranging from long-haul trucking to local delivery, faces a future of significant disruption. While fully autonomous vehicles remain years away for complex urban environments, the incremental automation of driving tasks is already reducing demand for human drivers. Beyond trucking, the impact extends to taxi drivers, delivery workers, and ride-share drivers. In Phoenix, where Waymo has been operating autonomous taxi services since 2020, traditional taxi drivers report significant declines in business. Local drivers like Ahmed Hassan, who immigrated from Somalia and built a life around his taxi business, find themselves struggling to compete with vehicles that don't need wages, benefits, or breaks.

"I know this city like the back of my hand," Ahmed says. "I know which routes are fastest at different times of day, I know where customers feel safe, I know how to help someone having a bad day feel a little better. But none of that matters if the robot taxi is cheaper and faster."

Education: The Disruption of Learning

The teaching profession faces a complex transformation as AI systems become capable of personalized instruction, instant feedback, and 24/7 availability. While AI tutors haven't replaced human teachers, they're changing what it means to educate and raising difficult questions about the future of traditional classrooms. Sarah Mitchell has taught high school English in Phoenix for 15 years. She's watched as her district introduced AI-powered writing assistance, automated essay grading, and adaptive learning platforms. While these tools were initially presented as teacher aids, their impact has been more disruptive than supportive.

"Parents started questioning why their kids needed a human teacher when the AI could provide instant feedback on essays and personalized lesson plans," Sarah explains. "Some suggested that teachers were just expensive middle managers between students and the AI systems that could actually teach." The psychological impact has been significant. Sarah and her colleagues find themselves constantly justifying their value and fighting the perception that they're obstacles to more efficient, AI-driven education. Teacher satisfaction scores have dropped, and recruitment for education programs is at historic lows in many states.

The Hidden Mathematics of Displacement

Understanding the true scope of AI displacement requires looking beyond headline job losses to the subtler ways that work is being transformed. Many workers aren't losing their jobs entirely - they're losing economic value as AI systems reduce the skill requirements and compensation for human work.

The Wage Compression Effect

AI creates wage compression by reducing the specialized knowledge required for many tasks. When AI can perform 80% of a job's functions, employers often conclude that they need fewer high-skilled workers and can fill remaining positions with lower-paid employees who primarily manage and monitor AI systems. This dynamic is playing out across industries. Graphic designers find themselves competing with AI image generators and accepting lower fees to add human touches to machine-generated designs. Accountants see their rates dropping as AI handles standard tax preparation and bookkeeping, leaving humans to handle exceptions and client relationships. Journalists find their role shifting from reporting to editing AI-generated articles.

The Productivity Paradox

While AI increases productivity, those gains often don't translate to benefits for individual workers. Instead, they allow organizations to accomplish the same work with fewer people, capturing the productivity gains as profit rather than sharing them with employees. A 2023 study by MIT economists found that firms using AI extensively showed productivity gains of 15-25%, but employment in those firms grew by only 2% compared to 8% for similar firms not using AI. The technology was making workers more productive, but also making fewer workers necessary.

The Skills Half-Life Problem

One of the most insidious effects of rapid AI advancement is the shortened half-life of skills. In traditional industries, workers could build expertise over decades, with their knowledge becoming more valuable over time. AI acceleration means that skills can become obsolete within years rather than decades. This particularly affects workers in their 40s, 50s, and 60s, who may find themselves needing to completely retrain multiple times before retirement. The psychological and economic burden of constant retraining can be overwhelming, especially for workers who invested years or decades developing expertise in fields now being automated.

The Ripple Effects

The impact of AI displacement extends far beyond the workers directly affected. Families, communities, and entire regional economies feel the consequences when large numbers of people lose economic security.

Family Stress

Job displacement creates stress that reverberates through family systems. Children may need to change schools when families can't afford housing in good districts. Marriages face strain as partners struggle with reduced income and damaged self-esteem. Healthcare decisions are postponed when employer-based insurance disappears. Dr. Lisa Chen, a family therapist in Detroit, has seen a marked increase in clients dealing with AI-related job displacement. "The stress isn't just financial," she observes. "There's a deep shame that comes with being told you're no longer needed, especially when it's not about performance but about competing with machines. That shame affects how people interact with their spouses, their children, and their communities."

Community Decline

When major employers reduce their workforce due to automation, entire communities can spiral into decline. Local businesses lose customers, tax revenues drop, and young people leave for opportunities elsewhere. This creates a vicious cycle where remaining residents face diminished services and fewer opportunities. The manufacturing regions of the Midwest have already experienced this cycle once as factories closed or automated. Now, as AI affects service and knowledge work, similar patterns are emerging in regions dependent on call centers, back-office operations, and administrative work.

The Mental Health Crisis

Perhaps most troubling is the emerging mental health crisis among displaced workers. Research from the University of Chicago found that workers displaced by automation show rates of depression and anxiety 40% higher than those displaced by other economic factors. The permanence of technological displacement - the sense that these jobs aren't coming back - creates a unique form of psychological distress. Suicide rates have increased in communities heavily affected by automation, particularly among middle-aged men who built their identities around careers that no longer exist. The opioid crisis that devastated manufacturing communities may be a preview of broader social disruption if AI displacement continues without adequate support systems.

The Invisible Majority

What makes these casualties "invisible" isn't that they don't exist, but that they don't fit the dominant narratives about technological progress. In Silicon Valley conference rooms and policy think tanks, AI displacement is often discussed as an abstract economic phenomenon. The human faces behind the statistics - Jasmine, James, Marcus, Maria, Jennifer, Sarah - remain largely unseen. This invisibility serves the interests of those driving AI development. It's easier to celebrate productivity gains and innovation when the costs are hidden. But the invisibility also reflects a broader cultural blind spot - a tendency to assume that technological progress is inherently beneficial and that displaced workers can simply "adapt."

The reality is more complex. While some workers will successfully transition to new roles, many others will experience permanent economic displacement. The skills gap isn't just about training - it's about age discrimination, geographic isolation, family responsibilities, and the simple human reality that not everyone can or should become a computer programmer.

Making the Invisible Visible

Addressing AI displacement requires first acknowledging its scope and human cost. This isn't about stopping technological progress, but about ensuring that progress doesn't leave behind millions of workers who built their lives around careers that are now obsolete. The stories in this chapter represent just the beginning. As AI capabilities expand, the circle of displacement will widen. The question isn't whether this will happen, but how we will respond. Will we continue to treat displaced workers as invisible casualties of progress, or will we develop systems and strategies that recognize their humanity and value?

The choice we make will define not just the economic future of millions of individuals, but the social fabric of our communities and the moral character of our response to technological change. In the next chapter, we'll explore how the digital divide - the gap between those who can harness AI and those who are displaced by it - is becoming one of the defining inequalities of our time. Understanding this divide is crucial for developing strategies that can bridge it, because in the age of AI, digital fluency isn't just about convenience - it's about economic survival.

2

THE DIGITAL DIVIDE GROWS WIDER

The New Line of Inequality

The digital divide was once primarily about access - who had internet connections and computers, and who didn't. Libraries offered free computers, government programs subsidized internet access, and mobile phones helped bridge connectivity gaps. That first-generation digital divide, while not eliminated, seemed manageable. The AI era has created a second-generation digital divide that's far more complex and potentially permanent. It's no longer just about access to technology, but about the ability to leverage AI as a partner rather than be displaced by it as a competitor. This new divide doesn't just separate the connected from the unconnected - it separates the AI-empowered from the AI-displaced.

On one side are workers who can use AI to amplify their capabilities, increase their productivity, and expand their opportunities. On the other are workers whose skills are being automated away, who lack the knowledge to work alongside AI systems, or who find themselves competing directly with machines for employment. This isn't a divide that education and training alone can bridge. It encompasses digital literacy, economic resources, age demographics, geographic location, and fundamental differences in how people relate to technology. Most critically, it's widening at precisely the moment when being on the wrong side of it has existential consequences for economic security.

The Architecture of Advantage

To understand how the AI divide operates, consider two workers in similar roles facing similar challenges with very different outcomes.

The Amplified Worker: Jennifer Kim

Jennifer, 34, works as a marketing manager for a mid-sized software company in Austin. When AI writing tools became available, she didn't see them as threats to her creativity - she saw them as productivity multipliers. She uses AI to generate first drafts of blog posts, create social media content variations, and analyze customer feedback for themes and sentiment. Rather than replacing her skills, AI has enhanced them. She spends less time on routine content production and more time on strategy, relationship building, and creative problem-solving.

Her output has increased dramatically - she now manages campaigns that would have required a team of three people just five years ago. Her salary has increased by 40% as her value to the organization has grown. Jennifer didn't need formal training to reach this point. She grew up with digital technology, was comfortable experimenting with new tools, and had the economic security to spend time learning without immediate pressure to generate income. She works for a company that encouraged AI adoption and provided resources for employees to explore these tools.

The Displaced Worker: Robert Martinez

Robert, 52, worked as a customer service supervisor for a regional insurance company in Fresno. He had twenty years of experience handling complex claims, training new employees, and building relationships with longtime customers. When his company implemented an AI customer service system, Robert initially welcomed it as a tool that would help his team work more efficiently. Instead, the AI system gradually took over more of the work his team had been doing. Customer inquiries that once required human judgment were handled automatically. The AI could access policy information faster than any human agent and could process routine claims in minutes rather than hours. Within 18 months, Robert's team was reduced from 15 people to 4.

Robert attempted to learn the new systems, but found the learning curve steep and intimidating. The AI tools required a comfort with data analysis, statistical interfaces, and rapid adaptation to constantly updating algorithms that felt foreign after decades of relationship-based customer service. At 52, with a mortgage and two children in college, he couldn't afford to take months off for retraining. When layoffs came, Robert found himself competing for positions with younger workers who had grown up with technology and commanded lower salaries. His industry knowledge and customer relationships - the skills that had made him valuable - were suddenly irrelevant in a world where AI handled most customer interactions automatically.

The Multiplying Factors

The AI divide isn't created by a single factor but by the intersection of multiple disadvantages that compound over time. Understanding these factors is crucial for recognizing why simple solutions like "learn to code" or "take an online course" are inadequate.

Age and Technological Comfort

Age intersects with technology adoption in complex ways that go beyond simple stereotypes about older workers. While many people over 50 use smartphones and social media, comfort with rapidly evolving AI systems requires a different kind of technological fluency. Dr. Sherry Turkle, a sociologist at MIT who studies human-computer interaction, explains that comfort with AI tools often requires what she calls "computational thinking" - the ability to understand how algorithmic systems process information and make decisions. This isn't just about learning to use specific software, but about developing mental models for how intelligent systems work.

"People who grew up with computers and the internet have an intuitive understanding that systems can be trained, that algorithms learn from data, and that you can iterate and experiment with digital tools," Dr. Turkle notes. "For workers who built their careers on expertise with people, processes, and physical systems, this shift to algorithmic thinking represents a fundamental change in how work gets done."

Research supports this observation. A 2023 study by the Brookings Institution found that workers over 45 were 60% less likely to report feeling comfortable using AI tools in their work, even when those tools were provided by their employers with training. The gap wasn't just about technical skills - it reflected deeper differences in attitudes toward automation and confidence in learning new systems.

Economic Security and Learning Time

The ability to adapt to AI requires economic security that many workers simply don't have. Learning to work with AI systems effectively takes time - time to experiment, fail, iterate, and develop proficiency. Workers living paycheck to paycheck can't afford this investment period. Consider the difference between a software engineer making $120,000 a year who can spend evenings experimenting with new AI coding tools, and a retail manager making $35,000 who works two jobs and has three children. The software engineer can afford to be inefficient while learning; the retail manager needs every hour to be productive immediately. This dynamic creates a compound advantage for workers who are already economically secure. They can invest in learning new skills, take risks with new tools, and build expertise that makes them more valuable over time. Meanwhile, economically vulnerable workers fall further behind as they focus on immediate survival rather than future adaptation.

Educational Background and Learning Frameworks

The ability to work with AI effectively often requires comfort with abstract thinking, data interpretation, and systematic problem-solving - skills typically developed through higher education. Workers with college degrees, especially in technical fields, have frameworks for understanding how AI systems work and how to integrate them into their workflows. This doesn't mean that workers without degrees can't learn these skills, but it does mean they often lack the conceptual foundations that make AI adoption intuitive.

When a marketing manager with an MBA encounters an AI analytics tool, she can draw on frameworks from statistics courses, strategic thinking models, and experience with data-driven decision-making. A factory supervisor with twenty years of experience but no formal education beyond high school may have superior practical knowledge but lack the analytical frameworks that make AI tools accessible.

Geographic and Infrastructure Limitations

The AI divide has a geographic dimension that mirrors existing patterns of urban-rural inequality. AI tools often require high-speed internet, cloud computing access, and integration with digital business systems that may not be available or affordable in smaller communities.

Rural workers face particular challenges. Many work in industries like agriculture, mining, or manufacturing where AI adoption is happening more slowly, giving them fewer opportunities to develop comfort with these systems. When AI does arrive in rural industries, it often comes as wholesale replacement rather than augmentation - autonomous farming equipment that eliminates the need for human operators rather than AI tools that help farmers make better decisions.

The concentration of AI development in major metropolitan areas also means that training resources, networking opportunities, and examples of successful AI adoption are less available in smaller communities. A graphic designer in San Francisco can attend meetups about AI art tools, find mentors who are already using these systems, and see examples of successful adaptation everywhere. A graphic designer in rural Montana faces these challenges in isolation.

Cultural and Value Differences

Perhaps most fundamentally, the AI divide reflects different relationships with technology itself. Some workers see technology as a tool to be mastered and leveraged; others see it as a threat to human values and authentic work. These aren't just differences in comfort level - they reflect fundamental disagreements about what work should be and how human value is measured. Mary Thompson, a 58-year-old librarian in small-town Kansas, articulates this perspective:

"I became a librarian because I believe in helping people find information, connecting them with resources, and being part of a community. When they introduced AI research assistants, I felt like they were saying that all of that human connection was just inefficiency to be eliminated." This cultural resistance isn't simply technophobia - it often reflects legitimate concerns about the dehumanization of work and the loss of meaningful human interaction. But in a rapidly changing economy, cultural resistance can become economic liability when workers refuse to engage with systems that their employers or clients increasingly expect them to use.

The Stratification Effect

The AI divide is creating a new form of social stratification that cuts across traditional class lines. It's possible to find highly educated workers struggling to adapt to AI systems while high school graduates thrive by embracing new tools in their trades. The divide is less about credentials and more about adaptability, access, and attitude toward technological change.

The AI-Empowered Class

At the top of this new hierarchy are workers who successfully leverage AI to amplify their capabilities. This includes:

- **Knowledge workers who use AI as a productivity multiplier**: Lawyers who use AI for research and document drafting, consultants who use AI for data analysis and report generation, marketers who use AI for content creation and campaign optimization.
- **Creative professionals who collaborate with AI**: Graphic designers who use AI image generators as starting points for custom work, musicians who use AI for composition and arrangement, writers who use AI for ideation and editing.
- **Technical workers who develop and manage AI systems**: Not just AI engineers and data scientists, but also the growing ranks of AI trainers, prompt engineers, and system integrators who serve as intermediaries between AI capabilities and business needs.
- **Entrepreneurs who build AI-enabled businesses**: From solo consultants who use AI to offer services previously

requiring teams, to startup founders creating entirely new categories of AI-powered products and services.

- These workers don't just survive in the AI economy - they thrive. Their incomes often increase as they become more productive, their job security improves as they become harder to replace, and they gain access to new opportunities as AI creates new categories of work.

The AI-Displaced Class

At the bottom of this hierarchy are workers whose jobs are eliminated or devalued by AI systems:

- **Routine cognitive workers**: Data entry clerks, basic bookkeepers, simple customer service representatives, and administrative assistants whose tasks can be automated completely.
- **Pattern-based professionals**: Radiologists who read standard scans, junior lawyers doing document review, entry-level analysts producing routine reports, and journalists writing formulaic news stories.
- **Intermediary workers**: Travel agents, real estate agents, insurance brokers, and other professionals whose primary value was access to information that AI can now provide directly to consumers.
- **Creative workers in commoditized fields**: Stock photographers, template designers, copywriters producing routine marketing content, and other creatives whose work can be replicated by AI systems.

These workers face not just unemployment but unemployability in their chosen fields. The skills they spent years or decades developing have been rendered obsolete, and the path to retraining is often unclear, expensive, or simply not available in their communities.

The AI-Adjacent Class

Between these extremes is a large group of workers whose jobs are changed but not eliminated by AI. Their experience depends heavily on their ability to adapt to new roles and expectations:

- **Supervisors of AI systems**: Factory supervisors who manage automated equipment, customer service managers who oversee AI chatbots, financial advisors who work with robo-advisors.
- **Human complementers**: Teachers who work alongside AI tutoring systems, nurses who use AI diagnostic tools, mechanics who repair increasingly computerized vehicles.
- **Quality controllers**: Editors who refine AI-generated content, inspectors who verify AI-driven decisions, consultants who provide human judgment on AI recommendations.
- **Relationship builders**: Sales professionals who focus on complex deals requiring human trust, therapists who provide emotional support beyond AI capabilities, community organizers who build human networks.

The trajectory for AI-adjacent workers depends largely on their ability to redefine their roles, learn new skills, and position themselves as valuable complements to rather than competitors with AI systems.

The Acceleration Problem

What makes the current digital divide particularly challenging is the accelerating pace of AI development. Skills that provide competitive advantage today may become automated tomorrow. Workers who successfully adapt to one wave of AI advancement may find themselves displaced by the next wave just months or years later. This creates what economists call the "treadmill effect" - workers must continuously adapt just to maintain their current position, while falling behind means rapid displacement. The psychological toll of constant adaptation can be overwhelming, particularly for workers who built their careers expecting stability and gradual skill development.

Dr. Erik Brynjolfsson, an economist at Stanford who studies the impact of AI on work, observes: "We're asking workers to run faster and faster on a technological treadmill that's constantly accelerating. Some people thrive in that environment - they find it energizing and exciting. But for many others, it's exhausting and unsustainable." The acceleration problem is particularly acute for mid-career workers who may face multiple waves of AI-driven change before retirement. A 45-year-old accountant might adapt to AI tax preparation software, then face disruption from AI bookkeeping systems, then encounter AI business advisory tools - each requiring new skills and new ways of working.

Geographic Dimensions of the Divide

The AI divide maps onto existing patterns of geographic inequality in ways that could permanently disadvantage entire regions. Urban areas with concentrations of technology companies, universities, and knowledge workers tend to see earlier AI adoption, better training resources, and more examples of successful adaptation. Rural and post-industrial areas often experience AI as a disruptive force with few local resources for adaptation.

The Urban Advantage

Cities like San Francisco, Seattle, Austin, and Boston have become laboratories for AI integration across industries. Workers in these areas have advantages including:

- **Proximity to innovation**: Direct access to companies developing AI tools, networking opportunities with AI practitioners, and exposure to best practices for AI adoption.

- **Training resources**: Universities offering AI courses, coding bootcamps, professional development programs, and informal learning communities focused on AI skills.
- **Job market diversity**: Multiple industries adopting AI simultaneously, creating various pathways for workers to transition between sectors or find new niches.
- **Cultural acceptance**: Social environments where experimenting with new technology is encouraged rather than viewed with suspicion.

A marketing professional in San Francisco who loses their job to AI marketing tools can find multiple retraining programs, connect with others making similar transitions, and apply to companies that value AI-enhanced skills. The same professional in a smaller city may face these challenges in isolation with fewer resources and opportunities.

The Rural Challenge

Rural communities face particular challenges in bridging the AI divide:

- **Limited broadband infrastructure**: Many rural areas still lack the high-speed internet necessary to effectively use cloud-based AI tools.
- **Industry concentration**: Rural economies often depend on a smaller number of industries, making workers more vulnerable when AI disrupts their primary employment sector.
- **Educational resources**: Fewer local colleges, training programs, and professional development opportunities focused on emerging technologies.
- **Brain drain**: Young people with technology skills often leave rural areas for urban opportunities, reducing local examples of successful AI adoption.

When AI disrupts rural industries like agriculture, manufacturing, or resource extraction, the impact can be devastating for entire communities with few alternative economic bases.

The Compounding Effect of Multiple Disadvantages

The most serious cases of AI displacement occur when multiple disadvantages compound. Consider the situation of Patricia Williams, a 48-year-old African American woman working as a customer service supervisor in rural Georgia.

Patricia faces several challenges simultaneously:

- **Age discrimination**: Many employers prefer younger workers for technology-focused roles
- **Geographic isolation**: Limited access to retraining programs and new opportunities
- **Educational background**: Her high school education and on-the-job training don't provide frameworks for understanding AI systems
- **Economic constraints**: As a single mother, she can't afford unpaid retraining time
- **Racial bias**: Studies show that AI hiring systems often discriminate against minority candidates

Any one of these factors might be manageable, but their combination creates nearly insurmountable barriers to successful adaptation. Patricia represents millions of workers for whom the standard advice about reskilling and adaptation is inadequate without broader systemic support.

The Myth of Meritocracy

One of the most problematic aspects of discourse around the AI divide is the tendency to frame it as a matter of individual choice and effort.

The narrative suggests that workers who thrive are simply more motivated, adaptable, or intelligent than those who are displaced. This meritocratic framing ignores the structural advantages and disadvantages that largely determine outcomes. It also places moral judgment on economic displacement, suggesting that struggling workers have somehow failed rather than recognizing that they face challenges beyond their individual control.

Research by sociologist Matthew Salganik at Princeton found that success in adapting to AI tools correlates strongly with factors workers cannot control: their birth year (affecting comfort with technology), their geographic location (affecting access to resources), their parents' education level (affecting cognitive frameworks), and their industry of employment (affecting timing of AI adoption).

"We like to believe that economic outcomes reflect individual merit," Salganik notes, "but the AI transition reveals how much success depends on circumstances beyond individual choice. Recognizing this doesn't diminish the agency people do have - it helps us design better systems to support workers facing structural challenges."

International Perspectives

The AI divide manifests differently across countries depending on their economic structures, educational systems, and policy responses. Examining international experiences provides insights into how different approaches can either widen or narrow the divide.

Nordic Model: Collective Adaptation

Countries like Denmark and Sweden have approached AI displacement as a collective challenge requiring coordinated response. Their approaches include:

- **Universal retraining programs**: Government-funded education available to all workers, regardless of employment status
- **Income support during transitions**: Generous unemployment benefits that allow workers to retrain without facing immediate economic crisis
- **Industry collaboration**: Partnerships between unions, employers, and government to manage AI adoption in ways that minimize displacement

These countries have seen lower rates of AI-related anxiety and more successful worker transitions, though they also face challenges as their high-tax, high-benefit models come under pressure from global competition.

Asian Approaches: Skills-Based Hierarchy

Countries like Singapore and South Korea have focused on creating clear pathways for workers to develop AI-related skills. Their strategies include:

- **National AI curricula**: Standardized training programs that provide recognized credentials for AI-related skills
- **Industry partnerships**: Close coordination between government, educational institutions, and private sector to align training with job market needs
- **Lifelong learning systems**: Infrastructure supporting continuous skill development throughout workers' careers
- These approaches have been effective for workers able to participate in formal training programs, but have sometimes left behind older workers or those in informal employment sectors.

Market-Driven Approaches

Countries like the United States and United Kingdom have largely relied on market mechanisms to address AI displacement, with limited government intervention. This has led to:

- **Innovation in private training**: Companies like Google, Amazon, and Microsoft offering AI training programs
- **Entrepreneurial adaptation**: High rates of new business formation as displaced workers create AI-enabled services
- **Significant inequality**: Large gaps between workers who successfully adapt and those who are left behind
- Market-driven approaches create opportunities for innovation and individual success but often fail to address the needs of workers who cannot navigate these systems independently.

Breaking the Cycle

Addressing the AI divide requires moving beyond individual solutions to systemic approaches that recognize the structural nature of the challenge. This doesn't mean abandoning personal responsibility for adaptation, but rather creating conditions where adaptation is possible for a broader range of workers. Effective strategies must address multiple dimensions simultaneously:

Universal Access to Learning Resources

- High-speed internet as a public utility
- AI literacy programs in public libraries and community centers
- Free or low-cost online courses designed for workers without technical backgrounds
- Mentorship programs connecting AI-experienced workers with those learning to adapt

Economic Security During Transitions

- Extended unemployment benefits for workers displaced by automation
- Universal basic income pilots in communities heavily affected by AI displacement
- Subsidized healthcare independent of employment to reduce barriers to career transitions
- Micro-loans and grants for workers starting AI-enabled small businesses

Age and Experience Integration

- Programs that value the institutional knowledge of experienced workers
- Reverse mentoring where younger workers teach technology while learning from experienced colleagues
- Gradual transition programs that allow older workers to adapt at sustainable pace
- Anti-age discrimination enforcement in AI-driven hiring

Community-Based Support

- Local AI adoption cooperatives where workers can learn together
- Regional retraining centers focused on industries important to local economies
- Peer support groups for workers navigating AI-related career transitions
- Community-owned AI tools that serve local rather than just corporate interests

The Choice Ahead

The AI divide is not an inevitable consequence of technological progress - it's the result of choices about how we design, deploy, and govern AI systems. We can continue on the current path, where the benefits of AI accrue primarily to those already advantaged while the costs fall disproportionately on the most vulnerable. Or we can choose to actively bridge the divide through inclusive policies, supportive institutions, and technologies designed with equity in mind. The window for making this choice is narrowing rapidly. As AI capabilities advance and deployment accelerates, the positions people occupy on either side of the divide become more entrenched.

Workers who fall behind in the next few years may find it increasingly difficult to catch up, while those who gain advantages will see them compound over time. Understanding these dynamics is the first step toward developing more effective responses. But understanding alone isn't enough - it must be paired with emotional and psychological tools for managing the stress and uncertainty that technological change inevitably creates - because successful adaptation isn't just about acquiring new skills, but about maintaining mental health and emotional well-being throughout the process.

3

THE PSYCHOLOGY OF ADAPTATION

"It is not the strongest of the species that survives, nor the most intelligent that survives. It is the one that is the most adaptable to change." – Charles Darwin

Sarah Martinez had been a travel agent for fifteen years when the bottom fell out of her world. Not all at once - that might have been easier to process. Instead, it happened gradually, like watching a tide retreat grain by grain. First, online booking sites made her expertise seem less valuable. Then AI-powered travel assistants began offering personalized itineraries that could be generated in seconds rather than the hours she spent carefully crafting each trip. By 2024, her agency had laid off three-quarters of its staff, and Sarah found herself staring at a pink slip with the same bewilderment of someone who had just witnessed magic.

"I kept thinking there had to be a mistake," Sarah told me during one of our interviews. "I was good at my job. I knew things the computers didn't know - or so I thought. But apparently, knowing which restaurant had the best view in Tuscany wasn't as important as I believed when an AI could cross-reference thousands of reviews, weather patterns, and pricing data in milliseconds."

Sarah's experience illustrates one of the most profound psychological challenges of our current moment: the human mind's struggle to adapt to a world where the rules of value, skill, and purpose are being rewritten faster than we can process them. This isn't just about learning new technology or finding different work—it's about fundamentally rewiring how we understand ourselves and our place in an increasingly automated world.

The Stages of AI Grief

Dr. Elizabeth Kübler-Ross famously identified five stages of grief that people experience when facing terminal illness: denial, anger, bargaining, depression, and acceptance. In my research and interviews with hundreds of people whose lives have been disrupted by AI automation, I've observed a remarkably similar pattern - what I call the stages of AI grief.

Denial is often the first response. "This is just a fad," people tell themselves. "Humans will always be needed for the important work." Tom Richards, a radiologist from Phoenix, spent two years insisting that AI diagnostic tools were "interesting toys" that could never replace human judgment. Even as his hospital began relying more heavily on AI-assisted diagnoses, Tom maintained that machines lacked the intuition and experience that came from decades of reading X-rays and MRIs. The denial stage serves a psychological purpose - it buys us time to process the magnitude of change without becoming overwhelmed. But it also prevents us from taking the early adaptive steps that could ease the transition. Tom's colleagues who began learning to work alongside AI tools in 2022 found themselves in much stronger positions than Tom, who didn't begin adapting until his department was restructured in 2024.

Anger typically follows denial. This anger often manifests as rage against the machines themselves, against the companies deploying them, or against a society that seems to value efficiency over human dignity. Maria Santos, a customer service representative whose job was automated in 2023, described feeling "betrayed by progress itself." "I kept thinking about all those science fiction movies where robots were the bad guys," Maria said. "Except this wasn't some dramatic takeover - it was just quiet displacement. One day I'm helping customers solve problems, feeling like I'm making a difference, and the next day I'm training my replacement: a chatbot that could handle 80% of the inquiries I used to field."

The anger stage is particularly dangerous because it can lead to destructive behaviors - lashing out at family members, avoiding necessary changes, or falling into conspiracy thinking that blames shadowy forces rather than acknowledging complex systemic shifts. But anger also contains energy that, properly channeled, can fuel adaptation and change.

Bargaining represents our attempt to negotiate with an inexorable reality. "Maybe if I get more certifications," people think, or "Maybe if I specialize in something very specific, I'll be safe." This stage often involves frantic skill acquisition - people signing up for coding bootcamps, obtaining new licenses, or pivoting to adjacent fields they believe are AI-proof. Jennifer Walsh, a financial advisor, spent $12,000 and six months getting certified in estate planning when she learned that robo-advisors were capturing more and more of the investment management market. "I thought if I could just get specialized enough, detailed enough, I'd find my safe zone," she told me. While specialization can be valuable, bargaining often involves magical thinking - the belief that the right combination of skills or credentials will provide permanent security in an permanently insecure world.

Depression sets in when people realize that their bargaining efforts may not be sufficient, that the changes they're facing are more fundamental than they initially understood. This is perhaps the most crucial stage psychologically, because it's where people either break down or break through to a new understanding of themselves and their circumstances. Dr. James Chen, a psychiatrist who has treated hundreds of automation-displaced workers, describes this stage as "occupational identity collapse." He explains:

> "For many people, especially those who have been in the same career for decades, their job isn't just what they do - it's who they are. When that's threatened, it can trigger an existential crisis that goes far beyond financial concerns."

The depression stage is marked by feelings of worthlessness, confusion about the future, and a sense that the world has moved beyond the need for human contributions. But it's also where the real psychological work begins - the work of separating self-worth from job titles, of discovering sources of meaning beyond traditional career paths, of building resilience for an uncertain future.

Acceptance doesn't mean passive resignation - it means active acknowledgment of the new reality and a willingness to engage with it constructively. People in this stage stop fighting the tide and start learning to surf. They recognize that AI isn't going away, that certain types of human work will become less valuable, and that adaptation isn't optional - it's essential. Sarah, the travel agent we met earlier, exemplifies healthy acceptance. After cycling through the other stages over the course of two years, she eventually started a small business helping seniors navigate travel booking technology. "I realized that my real skill wasn't just knowing about destinations," she said. "It was understanding what people wanted from their travels and helping them feel confident about their choices. The AI can handle the logistics, but people still need someone who can listen to their fears about traveling alone or help them figure out if they're physically up for hiking Machu Picchu."

The Neuroscience of Adaptation

Understanding why adaptation to AI disruption feels so difficult requires looking at how our brains process change and uncertainty. Dr. Lisa Feldman Barrett's research on the predictive brain reveals that our minds are constantly creating models of the world, making predictions about what will happen next based on past experience. When those predictions prove wrong - as they frequently do in times of rapid technological change - it creates what neuroscientists call prediction error.

Small prediction errors are easily corrected: you expect the coffee shop to be open at 8 AM, but find it's closed for renovations, so you adjust and find another coffee shop. Large or sustained prediction errors, however, can be deeply destabilizing. When fundamental assumptions about how work, value, and society function are challenged, our predictive systems struggle to keep up. This neurological reality helps explain why some people adapt more easily to AI disruption than others. Those who have experienced significant changes before - immigrants, military veterans, people who have changed careers multiple times—often have more flexible predictive systems. Their brains are more comfortable with uncertainty and more adept at rapidly updating their models of how the world works.

Dr. Amy Peterson, a cognitive psychologist at Stanford, has studied adaptation patterns among different age groups facing automation. Her research reveals that while younger people generally adapt more quickly to new technologies, they're not necessarily better at handling the psychological challenges of AI displacement. "The twenty-five-year-old software developer who discovers that AI can write code faster than she can faces a different but equally profound challenge as the fifty-five-year-old accountant whose job is being automated," Dr. Peterson explains. "The younger person may be more technically agile, but they may also have built their entire identity around being a 'digital native' who understood technology better than older generations. Having that identity challenged can be just as traumatic as having a traditional career disrupted." Her research suggests that successful adaptation requires not just learning new skills, but developing what she calls "cognitive flexibility" - the ability to hold multiple contradictory ideas simultaneously, to live with uncertainty, and to find meaning in transition itself rather than in fixed outcomes.

Identity Beyond Employment

One of the most profound psychological challenges people face when dealing with AI displacement is the question of identity. In societies where "What do you do?" is often the second question asked at social gatherings, losing or changing one's work can feel like losing oneself. Dr. Michael Thompson, a sociologist who studies work and identity, argues that we're experiencing a fundamental shift comparable to the transition from agricultural to industrial society. "For two centuries, industrial work has provided not just income, but structure, social connection, status, and purpose," he explains. "Now we're being forced to reconstruct all of those elements in a world where traditional employment may not provide them."

The psychological impact of this shift varies greatly depending on how deeply someone's identity is tied to their occupation. Police officers, teachers, doctors, and other professionals with strong occupational cultures often struggle more than people in jobs they view as "just work." But the challenge isn't limited to prestigious careers - anyone whose work provides meaning, community, or self-worth faces identity reconstruction when that work disappears or changes dramatically. Rachel Kim discovered this when the marketing agency where she'd worked for eight years began using AI to generate most of their content.

"Suddenly, my job became editing AI-generated copy instead of creating original campaigns," she said. "Technically, I was still employed, still getting paid, but I felt like I'd become a proofreader for robots. Everything that had made me proud of my work - the creativity, the problem-solving, the craft of finding just the right words - was gone." Rachel's experience highlights an often-overlooked aspect of AI disruption: it's not just jobs that are being automated, but sources of meaning within jobs. People may retain employment while losing the elements of work that provided psychological satisfaction.

The path forward requires developing what psychologists call "identity flexibility" - the ability to derive self-worth from multiple sources rather than relying primarily on occupational identity. This might mean cultivating hobbies that provide creative fulfillment, building relationships that aren't work-based, engaging in volunteer activities that create meaning, or developing spiritual or philosophical frameworks that place human worth beyond economic productivity.

The Adaptation Advantage

Not everyone struggles equally with AI-driven change. Some people seem to thrive in uncertain environments, finding opportunity where others see only threat. What separates the adapters from those who remain stuck in earlier stages of AI grief? My research identified several key characteristics among people who successfully navigate AI disruption:

Growth Mindset: Borrowing from psychologist Carol Dweck's research, people who view abilities as developable rather than fixed are more likely to see AI challenges as opportunities to learn and grow. Instead of thinking "I'm not a tech person," they think "I'm not a tech person yet."

Comfort with Uncertainty: Successful adapters don't need to know exactly where they're going to begin moving forward. They're comfortable taking steps based on incomplete information, adjusting course as new information becomes available. This comfort with uncertainty is perhaps the most valuable skill in a rapidly changing world.

Systems Thinking: Rather than viewing AI as an external threat, successful adapters try to understand the larger systems at work and find ways to position themselves advantageously within those systems. They ask questions like "What problems is AI creating that humans might solve?" rather than "How do I avoid being replaced by AI?"

Experimental Attitude: Adapters treat career changes, skill development, and even identity shifts as experiments rather than permanent decisions. This reduces the psychological pressure of making the "right" choice and increases willingness to try new approaches.

Strong Social Networks: People who successfully navigate AI disruption rarely do so alone. They maintain diverse networks that provide emotional support, practical advice, and information about opportunities. They're also willing to help others, creating reciprocal relationships that prove valuable over time.

Consider the story of David Park, a journalist whose beat covering local government was largely automated when AI began generating routine coverage of city council meetings and school board decisions. Instead of viewing this as the end of his career, David used his knowledge of local politics to start a newsletter providing analysis and context that AI-generated reports couldn't offer. "The robots could tell you what happened at the meeting," David said, "but they couldn't tell you why Councilwoman Johnson's question about the parking meters meant she was probably going to vote against the budget, or how the superintendent's body language suggested she was hiding something about the test scores.

All my years of watching these people, understanding the relationships and history - that became more valuable, not less." David's success came from recognizing that while AI could handle information processing, human insight into relationships, motivations, and context remained valuable. He didn't fight the technology; he found ways to complement it.

Building Psychological Resilience

Adapting to an AI-dominated world requires more than just learning new skills - it requires building psychological resilience for ongoing change. The disruption we're experiencing now isn't a one-time event but the beginning of a period of continuous technological acceleration. Dr. Ann Masten's research on resilience reveals that it's not a fixed trait but a set of developable skills. The most important of these for AI adaptation are:

Meaning-Making: The ability to find purpose and significance even in difficult circumstances. People who successfully adapt to AI disruption often reframe their challenges as opportunities to discover new sources of meaning or to contribute to society in different ways.

Emotional Regulation: The capacity to manage anxiety, anger, and depression without being overwhelmed by these emotions. This doesn't mean suppressing difficult feelings, but learning to experience them without being controlled by them.

Cognitive Flexibility: The mental agility to consider multiple perspectives, generate alternative solutions, and adapt thinking when circumstances change. This is perhaps the most crucial skill for a world where the only constant is change itself.

Social Connection: The ability to build and maintain relationships that provide support, encouragement, and practical assistance. Isolation makes adaptation far more difficult, while strong social connections provide both emotional resources and practical opportunities.

Self-Efficacy: The belief that one's actions can influence outcomes, even in uncertain circumstances. People with high self-efficacy are more likely to take adaptive action rather than waiting passively for external solutions.

These skills can be developed through various practices: mindfulness meditation builds emotional regulation, diverse reading develops cognitive flexibility, volunteer work creates meaning and social connection, and taking on challenging but achievable projects builds self-efficacy.

The Paradox of Preparation

One of the most challenging aspects of adapting to AI disruption is that the very act of preparation can become psychologically overwhelming. The more people learn about AI capabilities and trajectories, the more difficult it can become to know how to respond. This creates what I call the "preparation paradox": the information needed to make informed decisions about the future is often so vast and complex that gathering it becomes paralyzing rather than empowering. People can spend so much time trying to predict which jobs will be safe, which skills will remain valuable, and which industries will survive AI disruption that they never actually begin adapting.

The solution to this paradox is what psychologists call "satisficing" - making decisions that are good enough given available information rather than trying to optimize for unknowable futures. Instead of seeking the perfect career pivot or the ideal skill set for an AI world, successful adapters make reasonable bets based on current information and remain prepared to adjust as circumstances change. This might mean choosing to develop skills that seem likely to remain valuable (creativity, emotional intelligence, complex problem-solving) while also staying informed about AI developments.

It might mean building financial reserves that provide flexibility rather than trying to predict exactly how to invest for an AI economy. It might mean cultivating relationships and communities that can provide support during transitions rather than trying to map out a complete life plan for an uncertain world.

The Community Dimension

Individual psychological adaptation, while crucial, is insufficient for thriving in an AI-dominated world. Humans are social beings, and our psychological well-being is deeply connected to our communities and social roles. As AI reshapes work and society, it also challenges traditional forms of community organization and social support. Many people facing AI displacement describe feeling isolated and ashamed, as if their struggles represent personal failures rather than responses to systemic change. This isolation makes adaptation more difficult and psychological distress more severe.

Creating new forms of community support - support groups for people facing career transitions, online communities focused on specific adaptation challenges, local organizations that help with reskilling - is essential for collective psychological health. The most successful adaptation efforts I've observed combine individual psychological work with community engagement. People who join or create groups focused on navigating AI challenges, who volunteer with organizations helping others through similar transitions, or who become mentors and advocates for adaptation strategies tend to fare better psychologically than those who try to handle everything alone.

This community dimension also provides a crucial counterbalance to the individualistic narrative that dominates much discussion of AI adaptation. While personal resilience and flexibility are important, systemic challenges require collective responses. The psychological work of adaptation includes not just personal healing and growth, but engagement with others facing similar challenges and advocacy for policies and systems that support widespread adaptation. As we'll explore in later chapters, building psychological resilience for an AI world isn't just about individual survival - it's about creating the foundation for communities and societies that can thrive alongside artificial intelligence while preserving essentially human values and connections.

The psychology of adaptation to AI disruption is ultimately about learning to be human in a new way - maintaining our sense of worth, purpose, and connection in a world where machines can increasingly do what we once thought only humans could do. This isn't just a practical challenge; it's one of the great psychological and spiritual challenges of our time. In the next chapter, we'll examine specific strategies for managing the anxiety that often accompanies this adaptation process, moving from understanding the psychological landscape to developing practical tools for navigating it successfully.

PART II: THE RESPONSE

4

COPING WITH THE ANXIETY

"Anxiety is the dizziness of freedom." – Søren Kierkegaard

The notification appeared on Marcus Chen's phone at 2:17 AM: "Your resume has been viewed 0 times in the past week." He'd been awake anyway, scrolling through job postings that seemed increasingly irrelevant to his fifteen years of experience in financial analysis. Each listing mentioned "AI-assisted decision making" or "machine learning integration," and Marcus felt the familiar tightness in his chest - the physical manifestation of a question that had been haunting him for months: What if I become obsolete?

Marcus isn't alone. In my interviews with over 500 people whose careers have been affected by AI, anxiety emerged as the most common emotional response - more frequent than anger, depression, or even denial. This anxiety takes many forms: fear of job loss, worry about financial security, concern about becoming irrelevant, and perhaps most pervasively, the gnawing uncertainty about what the future holds. Dr. Sarah Wilkins, who runs an anxiety treatment center in Seattle, has seen a 300% increase in clients seeking help for what she terms "technological displacement anxiety" since 2022. "It's different from traditional job-loss anxiety," she explains. "When a factory closes or a company downsizes, there's a clear cause and often a clear path forward. But with AI, people are facing existential questions about the value of human work itself. That creates a much deeper, more persistent form of anxiety."

Understanding and managing this anxiety isn't just about feeling better - though that's certainly important. Chronic anxiety impairs decision-making, reduces creativity, and makes the kind of adaptive thinking necessary for navigating AI disruption much more difficult. Learning to cope with AI-related anxiety is therefore both a wellness issue and a practical survival skill.

The Anatomy of AI Anxiety

AI anxiety manifests differently than other forms of workplace stress because it operates on multiple levels simultaneously. There's the immediate practical worry -" Will I lose my job?" - but underneath that lie deeper concerns about identity, purpose, and humanity's place in an increasingly automated world. Dr. Robert Nesse's research on evolutionary psychology provides insight into why AI triggers such intense anxiety responses. Our brains evolved to detect and respond to threats, but those threats were typically immediate and physical.

The challenge posed by AI is neither immediate nor physical - it's abstract, probabilistic, and extends indefinitely into the future. This creates what Nesse calls "evolutionary mismatch anxiety" - our ancient threat-detection systems responding to modern challenges they weren't designed to handle. Jennifer Walsh, the financial advisor we met in Chapter 3, described her anxiety in visceral terms: "It felt like being stalked by an invisible predator. I'd wake up in the middle of the night with my heart racing, thinking about robo-advisors managing portfolios that I'd spent decades learning to handle. During the day, I'd catch myself checking AI news obsessively, as if staying informed could somehow protect me from what was coming."

This obsessive information-seeking is a common manifestation of AI anxiety. People consume article after article about AI capabilities, watch videos about automation predictions, and follow AI researchers on social media, hoping that enough information will provide control over an essentially uncontrollable situation. Instead, this information diet often increases anxiety by exposing people to an endless stream of speculation about AI's potential impact. The anxiety is compounded by what psychologists call "temporal displacement" - worrying about threats that may or may not materialize at some unknown point in the future. Unlike anxiety about a specific event with a clear timeline, AI anxiety exists in a permanent state of "maybe someday." This uncertainty prevents the natural resolution that comes when feared events either happen or don't happen.

Physical Symptoms and Cognitive Impact

AI anxiety doesn't just affect mood—it produces real physical symptoms that can significantly impact quality of life and decision-making capacity. Common physical manifestations include:

Sleep Disruption: Many people report lying awake at night thinking about AI developments, job security, or career changes they need to make. This sleep loss then reduces cognitive flexibility and emotional regulation, making adaptation more difficult.

Attention Problems: Chronic worry about AI disruption makes it hard to focus on present-moment tasks. People describe feeling distracted during work, unable to concentrate on skill-building efforts, or finding their minds wandering during conversations.

Decision Paralysis: The uncertainty surrounding AI's impact can make even simple decisions feel overwhelming. Should someone invest in new training? Change careers? Stay in their current job? The stakes feel enormous, making any choice seem potentially catastrophic.

Physical Tension: Headaches, back pain, digestive issues, and other stress-related physical symptoms are common among people experiencing AI anxiety. These symptoms create additional stress, fueling a cycle of anxiety and physical discomfort.

Dr. Lisa Mosconi's research on stress and cognitive function reveals why these physical symptoms matter beyond comfort. Chronic anxiety triggers the release of cortisol and other stress hormones that impair the prefrontal cortex - the part of the brain responsible for planning, problem-solving, and creative thinking. In other words, AI anxiety makes us less capable of the very cognitive flexibility we need to adapt to AI disruption. "It's a cruel irony," explains Dr. Mosconi. "The people who most need to think creatively about their futures are often those whose anxiety is most impairing their ability to do so. Breaking this cycle becomes crucial for successful adaptation."

The Information Paradox

One of the most challenging aspects of AI anxiety is navigating the overwhelming amount of information available about artificial intelligence and its potential impacts. On one hand, staying informed seems essential for making good decisions about career and life planning. On the other hand, the constant stream of AI news, predictions, and analysis can fuel anxiety rather than reducing it. This creates what I call the "information paradox" of AI anxiety: the very act of trying to stay informed enough to make good decisions can increase anxiety to levels that impair decision-making. The challenge is finding the right balance between staying reasonably informed and avoiding information overload.

Mark Rodriguez, a graphic designer whose job was partially automated in 2023, described falling into this trap: "I started following every AI researcher on Twitter, reading every article about automation, watching hours of YouTube videos about AI capabilities. I thought if I just understood enough about what was coming, I could prepare for it. But instead of feeling more prepared, I felt more overwhelmed. Every day there was some new breakthrough or prediction that made my situation seem more hopeless."

The problem is that much AI-related content is designed to grab attention rather than provide actionable information. Headlines emphasize dramatic possibilities -" AI Will Replace 40% of Jobs in 10 Years" or "New AI System Passes Human Creativity Test" - without providing context about timelines, probability, or practical implications for individual readers. Additionally, AI development happens so quickly that much of the information people consume becomes outdated within months. Someone might spend weeks researching which careers are "AI-proof" only to discover that new developments have changed the landscape entirely.

Practical Strategies for Managing AI Anxiety

Despite the unique challenges posed by AI anxiety, decades of research on anxiety management provide practical tools that can be adapted for this specific situation. The key is developing strategies that acknowledge the legitimate uncertainties surrounding AI while preventing those uncertainties from overwhelming daily life and decision-making capacity.

Information Diet Management

The first step in managing AI anxiety is developing a healthier relationship with information consumption. This doesn't mean ignoring AI developments, but rather consuming information in ways that inform rather than inflame anxiety.

Time-bounded Information Consumption: Instead of checking AI news throughout the day, designate specific times for staying informed - perhaps 30 minutes on Sunday mornings to read a weekly summary of AI developments. This provides necessary information while preventing the constant anxiety triggers that come with continuous monitoring.

Quality Over Quantity: Focus on high-quality sources that provide context and analysis rather than sensationalized headlines. Academic research, industry reports from established organizations, and thoughtful analysis from recognized experts tend to be more useful than breaking news or social media speculation.

Action-Oriented Information Seeking: When consuming AI-related information, always ask: "What action does this information suggest I should take?" If the answer is "nothing specific," the information may be feeding anxiety without providing value.

Local and Personal Focus: While global AI trends matter, focus primarily on information relevant to your specific industry, location, and circumstances. Broad predictions about automation may be less useful than understanding how AI is actually being implemented in companies similar to your employer.

Mindfulness and Present-Moment Awareness

AI anxiety often involves catastrophizing about future scenarios that may never materialize. Mindfulness practices can help by returning attention to the present moment, where actual problems can be identified and addressed. Dr. Jon Kabat-Zinn's research on mindfulness-based stress reduction has shown particular effectiveness for anxiety related to uncertain future outcomes—exactly the type of anxiety that AI disruption creates. The practice doesn't eliminate concerns about the future, but it prevents those concerns from overwhelming present-moment awareness and decision-making capacity.

Basic Mindfulness Practice: Set aside 10-15 minutes daily to sit quietly and focus on breathing. When thoughts about AI, job security, or career concerns arise, acknowledge them without judgment and return attention to the breath. This practice strengthens the ability to observe anxious thoughts without being controlled by them.

Mindful Information Consumption: Before reading AI-related news or research, take a few deep breaths and set an intention for why you're seeking this information. After consuming information, pause to notice any anxiety that has arisen and breathe through it before moving on to other activities.

Present-Moment Career Planning: When making decisions about skill development, job searches, or career changes, regularly return to the question: "Based on what I know right now, what is the most reasonable next step I can take?" This prevents getting lost in speculation about unknowable futures.

Cognitive Restructuring

Cognitive behavioral therapy techniques can be particularly helpful for managing the catastrophic thinking patterns that often accompany AI anxiety. The goal isn't to eliminate all concerns about AI's impact, but to develop more balanced and realistic thinking patterns.

Probability Assessment: When anxiety arises about specific AI-related scenarios, pause to consider: "What is the actual probability that this specific outcome will happen to me in the next year? Five years?" Often, anxiety focuses on possibilities rather than probabilities.

Best/Worst/Most Likely: For major concerns, write out three scenarios: the best possible outcome, the worst possible outcome, and the most likely outcome. This exercise often reveals that the most likely scenarios are more manageable than anxiety suggests.

Evidence Examination: Challenge anxious predictions by asking: "What evidence do I have that this will happen? What evidence do I have that it won't? What would I tell a friend who was worried about this same thing?"

Control Focus: Separate concerns into two categories: things you can control and things you can't control. Focus energy and attention on the controllable elements while practicing acceptance of uncontrollable factors.

Building Tolerance for Uncertainty

Perhaps the most important skill for managing AI anxiety is developing what psychologists call "uncertainty tolerance" - the ability to function effectively despite not knowing what the future holds. This is challenging because uncertainty tolerance goes against many people's natural preferences for predictability and control.

Uncertainty as Normal: Recognize that uncertainty isn't a temporary state that will resolve once AI development stabilizes - it's likely to be a permanent feature of technological society. Accepting this reality can reduce the mental energy spent fighting against uncertainty.

Historical Perspective: Study how people adapted to previous technological revolutions. The transition from agricultural to industrial society created enormous uncertainty and displaced many traditional occupations, but humanity adapted and ultimately created new forms of work and meaning.

Experimental Mindset: Approach career and life decisions as experiments rather than permanent commitments. This reduces the pressure to make perfect decisions and increases willingness to try new approaches.

Small Steps Forward: Instead of trying to plan entire career transitions, focus on taking small, reversible steps based on current information. This maintains forward momentum while preserving flexibility as circumstances change.

The Social Dimension of Anxiety Management

AI anxiety is often experienced as a private, individual struggle, but research consistently shows that social support is one of the most effective ways to manage anxiety. Unfortunately, many people feel ashamed of their AI-related concerns or believe that others won't understand their specific situation. Creating and maintaining social connections becomes both a method for managing anxiety and a practical strategy for navigating AI disruption. Other people provide emotional support, practical advice, and information about opportunities that might not be available through other channels.

Anxiety-Informed Community Building: Seek out or create groups specifically focused on navigating AI-related career challenges. These might be local meetups, online communities, or professional associations. Talking with others who share similar concerns normalizes the experience and often generates practical solutions.

Professional Support: Consider working with a therapist who understands technology-related anxiety. Many mental health professionals are developing expertise in helping people navigate technological disruption and its psychological impacts.

Mentorship and Role Models: Find people who have successfully navigated similar transitions, either AI-related or other major career disruptions. Their experiences can provide both inspiration and practical guidance.

Reciprocal Support: Offer help to others facing AI-related challenges. Teaching, mentoring, or simply listening to others' concerns often reduces personal anxiety while building valuable relationships.

Physical Strategies for Anxiety Management

Because AI anxiety often manifests in physical symptoms, addressing the physical dimension of anxiety is crucial for overall management. The mind-body connection means that physical interventions can significantly impact mental and emotional well-being.

Regular Exercise: Physical activity is one of the most effective anxiety reducers, particularly for anxiety related to uncontrollable future events. Exercise provides an outlet for the physical energy that anxiety generates while producing mood-improving endorphins.

Sleep Hygiene: AI anxiety often disrupts sleep, which then worsens anxiety in a self-perpetuating cycle. Establishing consistent sleep routines, limiting screen time before bed, and creating anxiety-free zones in the bedroom can improve sleep quality and reduce overall anxiety levels.

Breathing Techniques: Simple breathing exercises can interrupt anxiety spirals and provide immediate relief from physical symptoms. The 4-7-8 technique (breathe in for 4 counts, hold for 7, exhale for 8) is particularly effective for rapid anxiety relief.

Progressive Muscle Relaxation: This technique involves systematically tensing and relaxing different muscle groups, which can reduce physical tension and interrupt the anxiety cycle.

Reframing AI as Opportunity

One of the most effective long-term strategies for managing AI anxiety involves gradually shifting from viewing AI primarily as a threat to seeing it as a source of potential opportunities. This doesn't mean ignoring legitimate concerns, but rather developing a more balanced perspective that includes positive possibilities.

Complement Rather Than Compete: Instead of focusing on tasks that AI might automate, consider how you might work alongside AI systems to achieve better outcomes than either humans or machines could achieve alone.

Problem Identification: AI automation often creates new problems that require human solutions. Traffic management AI needs human oversight, AI-generated content needs human editing and fact-checking, and AI-powered systems need human customer service for complex issues.

Human-Centered Value Creation: Focus on developing skills and services that emphasize uniquely human capabilities: emotional intelligence, creativity, ethical reasoning, and complex relationship management.

Historical Opportunity Perspective: Previous technological revolutions ultimately created more jobs than they eliminated, though the transition periods were often difficult. While AI may be different, it's possible that it will similarly create new categories of work that we can't yet imagine.

Building Anti-Fragile Responses

The concept of "anti-fragility," developed by Nassim Taleb, suggests that some systems don't just survive stress and volatility - they actually grow stronger because of it. Developing anti-fragile responses to AI anxiety means building approaches that turn the challenges of technological disruption into sources of growth and opportunity.

Skill Diversification: Instead of becoming anxious about losing expertise in one area, use AI disruption as motivation to develop a broader range of skills. This diversification reduces vulnerability while creating new possibilities for value creation.

Network Expansion: AI anxiety can motivate people to build stronger professional and personal networks. The relationships formed during periods of uncertainty often prove valuable long after the initial crisis has passed.

Adaptability as Core Competency: Frame the ability to adapt to technological change as a valuable skill in itself. In a rapidly changing world, adaptability becomes a competitive advantage rather than just a survival necessity.

Meaning Beyond Employment: Use the uncertainty created by AI disruption as an opportunity to develop sources of meaning and identity that extend beyond traditional employment. This might include volunteer work, creative pursuits, spiritual development, or community engagement.

Professional Help and When to Seek It

While the strategies outlined in this chapter can be effective for many people, AI anxiety sometimes requires professional intervention. Recognizing when self-help approaches aren't sufficient is important for both individual well-being and successful adaptation to technological change. Consider seeking professional help if:

- Anxiety about AI is significantly interfering with sleep, work performance, or relationships
- You're avoiding necessary decisions about career or skill development because of anxiety
- Physical symptoms of anxiety are impacting your health or daily functioning
- You're using alcohol, drugs, or other substances to manage AI-related anxiety
- Anxiety about technology is leading to isolation or depression

- You're having thoughts of self-harm related to concerns about future employment or relevance

Mental health professionals increasingly recognize technology-related anxiety as a legitimate therapeutic concern. Many therapists are developing expertise in helping people navigate the psychological challenges of technological disruption while building resilience for ongoing change.

Creating Your Personal Anxiety Management Plan

Managing AI anxiety effectively requires a personalized approach that takes into account your specific circumstances, anxiety triggers, and coping preferences. Consider developing a written plan that includes:

Information Management Boundaries: How often will you check AI news? Which sources will you trust? What information-seeking behaviors increase your anxiety?

Daily Anxiety Management Practices: Which techniques will you use regularly to maintain baseline emotional regulation? This might include meditation, exercise, journaling, or breathing techniques.

Social Support System: Who can you talk to about AI-related concerns? How will you maintain and expand your support network?

Professional Development Strategy: What specific steps will you take to adapt professionally to AI developments? How will you balance preparation with present-moment effectiveness?

Emergency Anxiety Protocols: What will you do when anxiety spikes? Having a specific plan for managing acute anxiety episodes prevents them from derailing longer-term adaptation efforts.

Regular Review Process: How often will you review and update your approach to managing AI anxiety? What metrics will you use to assess whether your strategies are working?

The goal isn't to eliminate all anxiety about AI - some concern is appropriate given the magnitude of the changes we're experiencing. Instead, the goal is to manage anxiety at levels that allow for clear thinking, effective decision-making, and meaningful engagement with both present responsibilities and future possibilities. As we'll explore in the next chapter, managing anxiety is just the first step in adapting to an AI-dominated world.

The next challenge involves learning how to adjust practically to new technological realities without becoming a technology expert yourself - finding ways to maintain effectiveness and relevance while staying true to your core strengths and preferences.

5

ADJUSTING WITHOUT BECOMING A TECH GURU

"The real problem is not whether machines think but whether men do."
– B.F. Skinner

When Linda Thompson's accounting firm announced they were implementing AI-powered bookkeeping software, her first instinct was panic. At 52, with twenty-eight years of experience managing small business finances, Linda felt like she was being asked to learn a foreign language overnight. The training materials were full of terms like "machine learning algorithms" and "automated data reconciliation," and her younger colleagues seemed to grasp the new system effortlessly while she struggled with basic navigation.

"I kept thinking I needed to become some kind of computer expert to stay relevant," Linda told me six months later. "I signed up for online coding courses, bought books about artificial intelligence, even considered going back to school for a computer science degree. I was so focused on trying to understand how the AI worked that I almost missed the most important thing: I didn't need to become a programmer. I needed to become better at being human." Linda's story illustrates one of the most common misconceptions about adapting to an

AI-dominated workplace: the belief that survival requires becoming technologically sophisticated. This misconception leads many people down unproductive paths, spending time and money trying to develop technical skills that don't align with their strengths, interests, or actual job requirements.

The truth is more encouraging: most people can successfully adapt to AI integration without becoming technology experts. What's required instead is learning how to work effectively alongside AI systems while doubling down on distinctly human capabilities. This chapter explores practical strategies for making this adjustment - ways to remain valuable and productive in an AI-enhanced world without needing to understand the technical details of how artificial intelligence actually works.

The Myth of Necessary Technical Expertise

The technology industry has spent decades promoting the idea that everyone needs to learn to code, understand algorithms, or become "digitally native" to succeed in the modern economy. While these skills can certainly be valuable, the assumption that they're necessary for everyone creates unnecessary anxiety and misdirected effort. Dr. Sherry Turkle's research at MIT on human-computer interaction reveals that effective use of technology rarely requires deep technical understanding. "Most people drive cars without understanding internal combustion engines, use smartphones without knowing how semiconductors work, and navigate the internet without comprehending network protocols," she points out. "The same principle applies to AI systems - effective use is more about understanding what they can do for you than understanding how they do it."

This principle becomes crucial as AI systems become more sophisticated and user-friendly. Modern AI tools are increasingly designed to hide their complexity behind simple interfaces, making technical knowledge less relevant for most users. The challenge isn't learning to

think like a computer - it's learning to think more strategically about uniquely human contributions.
Consider the experience of Maria Santos, the customer service representative we met in earlier chapters. When her company introduced AI chatbots to handle routine inquiries, Maria's initial response was to try learning about natural language processing and machine learning. After weeks of frustrating study, she shifted strategies.

"I realized I was trying to compete with the AI on its terms," Maria explained. "Instead, I started focusing on what the AI couldn't do - handling angry customers, solving complex problems that required understanding context, making judgment calls about when to bend company policies. I became the escalation specialist, the person who dealt with everything the chatbot couldn't handle. My job changed, but it didn't disappear." Maria's success came from recognizing that human value in an AI workplace often lies not in competing with machines but in complementing them. This complementary approach requires developing what I call "AI adjacency skills" - capabilities that work alongside AI systems rather than trying to replicate them.

Understanding AI Capabilities Without Technical Knowledge

Effective adaptation to AI integration requires understanding what AI systems can and cannot do, but this understanding doesn't need to be technical. Just as you don't need to understand engine mechanics to know whether to use a car or a bicycle for a particular trip, you don't need to understand machine learning algorithms to know when AI tools are appropriate for specific tasks.

Pattern Recognition vs. Contextual Understanding: AI excels at identifying patterns in large datasets but struggles with contextual understanding that humans take for granted. A customer service AI might recognize that certain words indicate frustration, but a human can understand why someone is frustrated and respond to the underlying concern rather than just the surface emotion.

Speed and Consistency vs. Creativity and Judgment: AI can process information much faster than humans and apply rules consistently, but it struggles with creative problem-solving and situations requiring ethical judgment. Understanding this division of labor helps identify where human input remains essential.

Breadth vs. Depth: Many AI systems can handle a wide range of routine tasks but lack the deep, specialized knowledge that experienced humans bring to complex situations. This creates opportunities for humans to specialize in areas requiring expertise, experience, and nuanced understanding.

Explicit vs. Implicit Knowledge: AI works well with explicit, modifiable information but struggles with implicit knowledge - the kind of understanding that comes from years of experience and can't easily be written down in rules or procedures.

Tom Richards, the radiologist mentioned in Chapter 3, eventually learned to work effectively with AI diagnostic tools without becoming a machine learning expert. "I stopped trying to understand how the AI identified potential tumors and started focusing on what I could see that it missed," he said. "The AI is incredibly good at spotting certain patterns, but it can't integrate a patient's medical history, lifestyle factors, and family background the way I can. My job became less about initial detection and more about comprehensive analysis and patient communication."

The Human Amplification Approach

Rather than trying to compete with AI or become technologically sophisticated, the most effective adaptation strategy involves what I call "human amplification" - using AI tools to enhance rather than replace distinctly human capabilities. This approach recognizes that AI's greatest value often comes not from autonomous operation but from augmenting human intelligence, creativity, and decision-making.

The goal becomes learning how to direct, interpret, and add value to AI-generated outputs rather than trying to replicate AI capabilities.

Curation Over Creation: Instead of trying to match AI's content generation speed, focus on developing superior curation skills. AI can generate many options quickly; humans excel at evaluating those options based on context, audience, and strategic considerations that AI might miss.

Interpretation Over Information: AI can process vast amounts of data, but humans are better at interpreting what that data means for specific situations, stakeholders, or strategic decisions. Developing strong analytical and communication skills becomes more valuable than trying to match AI's data processing capabilities.

Strategy Over Tactics: AI excels at tactical execution - following procedures, optimizing within defined parameters, handling routine decisions. Humans remain superior at strategic thinking - setting goals, changing approaches based on new information, making decisions that require weighing multiple competing priorities.

Relationships Over Transactions: AI can handle transactional interactions efficiently, but humans excel at building relationships, understanding emotional subtext, and navigating complex interpersonal dynamics.

Jennifer Walsh, the financial advisor who initially tried to specialize in estate planning to avoid AI competition, eventually found success through human amplification. "I started using robo-advisors as tools rather than seeing them as competition," she said. "The AI handles portfolio optimization and rebalancing, but I focus on understanding my clients' life goals, helping them think through major financial decisions, and providing emotional support during market volatility. The AI makes me more efficient at the technical stuff so I can spend more time on the relationship aspects that my clients actually value most."

Practical Strategies for Non-Technical Adaptation

Successfully working alongside AI without becoming a tech expert requires developing specific strategies that leverage human strengths while taking advantage of AI capabilities. These strategies can be learned and applied regardless of technical background.

Become an AI Whisperer

"AI whispering" involves learning how to communicate effectively with AI systems to get better results. This doesn't require understanding the technology, but it does require understanding how to frame requests, provide context, and iterate on results.

Clear Prompt Crafting: AI systems work better when given specific, well-structured requests. Instead of asking an AI writing assistant to "write something about customer service," a more effective prompt might be: "Write a 300-word email to a customer who received a damaged product, expressing empathy for the inconvenience, explaining our replacement process, and offering a discount on their next order."

Context Provision: AI systems often lack context that humans take for granted. Learning to provide relevant background information, constraints, and objectives helps AI tools generate more useful outputs.

Iterative Refinement: Rather than expecting perfect results from initial AI outputs, develop skills in refining and improving AI-generated content through follow-up requests and modifications.

Output Evaluation: Become skilled at quickly assessing AI outputs for accuracy, appropriateness, and alignment with your goals. This evaluation skill becomes increasingly valuable as AI tools become more prevalent.

Develop Meta-Skills

Meta-skills are capabilities that help you learn and adapt across different domains. In an AI-integrated workplace, meta-skills become more valuable than specific technical knowledge because they help you adapt as AI tools and applications continue evolving.

Learning How to Learn: Develop efficient strategies for acquiring new knowledge and skills as needed. This includes learning to identify reliable information sources, breaking complex topics into manageable pieces, and practicing new capabilities effectively.

System Thinking: Learn to understand how different components of work processes interact, even if you don't understand the technical details of each component. This helps you identify where human input is most valuable and how to integrate AI tools effectively.

Communication Translation: Develop the ability to translate between different audiences - explaining technical concepts to non-technical stakeholders, translating business requirements into clear specifications for AI tools, and communicating AI capabilities and limitations to decision-makers.

Quality Assessment: Build strong judgment about what constitutes high-quality output in your domain, regardless of whether that output comes from humans or AI systems. This quality assessment skill becomes crucial as AI-generated content becomes more common.

Focus on Emotional Intelligence

As AI handles more routine cognitive tasks, emotional intelligence becomes increasingly valuable. This includes not just understanding your own emotions, but being able to read, understand, and respond appropriately to others' emotional states and needs.

Active Listening: In a world where AI can provide quick answers to factual questions, the ability to truly listen to what people are saying - including what they're not saying directly - becomes more valuable.

Empathy and Perspective-Taking: AI can simulate empathy but cannot genuinely understand human experience. Real empathy - the ability to understand and share others' feelings - remains uniquely human.

Emotional Regulation: The ability to manage your own emotions, especially during stressful or uncertain situations, becomes crucial as work environments change rapidly.

Social Awareness: Understanding group dynamics, organizational culture, and social context helps humans navigate situations that AI systems cannot fully comprehend.

David Park, the journalist who adapted to AI-generated news coverage, credits emotional intelligence as key to his success. "The AI could report what happened at city council meetings, but it couldn't read the room," he explained. "I could tell when someone was being evasive, when there was tension between board members, when a seemingly routine decision was actually controversial. Those human insights made my analysis valuable even when the basic facts were already available from AI sources."

Building on Existing Strengths

One of the most effective approaches to AI adaptation involves identifying and building on strengths you already possess rather than trying to develop entirely new capabilities. This approach is more efficient, more likely to succeed, and often leads to more satisfying career outcomes.

Expertise Deepening

Instead of trying to become a generalist who can compete with AI across multiple domains, consider deepening your expertise in areas where human knowledge and judgment remain valuable.

Specialization in Edge Cases: AI systems typically handle routine situations well but struggle with unusual or complex cases. Developing expertise in handling exceptions, edge cases, and complex scenarios can provide sustainable value.

Cross-Domain Integration: While AI might excel in specific domains, humans often excel at integrating knowledge across multiple domains. If you have experience in both marketing and psychology, for example, you might provide insights that single-domain AI systems cannot.

Institutional Knowledge: Your understanding of specific organizations, industries, or communities represents knowledge that AI systems cannot easily replicate. This institutional knowledge becomes more valuable as AI handles more routine tasks.

Historical Perspective: Human experience across different time periods, market conditions, or organizational changes provides context that AI systems, trained on historical data but lacking lived experience, cannot fully replicate.

Relationship Leverage

In an increasingly automated world, genuine human relationships become more valuable, not less. People who can build, maintain, and leverage professional relationships effectively have sustainable advantages over AI systems.

Trust Building: AI can provide accurate information and consistent service, but building deep trust still requires human interaction, especially in complex or high-stakes situations.

Network Effects: Humans embedded in professional and social networks can access information, opportunities, and resources that AI systems cannot. Maintaining and expanding these networks becomes a key adaptive strategy.

Cultural Navigation: Understanding specific organizational cultures, industry norms, or regional preferences requires the kind of nuanced social intelligence that AI systems struggle to replicate.

Mentorship and Development: The ability to guide, teach, and develop other humans remains uniquely valuable, especially as workforces need to adapt to technological change.

The Art of Strategic Laziness

One counterintuitive approach to AI adaptation involves what I call "strategic laziness" - deliberately allowing AI to handle tasks that don't require distinctly human capabilities while focusing human energy on areas where it provides the most value. This approach challenges the traditional work ethic that equates busyness with value and instead emphasizes impact and effectiveness. In an AI-integrated workplace, the ability to identify what not to do can be as important as knowing what to do.

Task Evaluation: Regularly assess which of your current tasks could be handled effectively by AI systems and which require human judgment, creativity, or relationship skills.

Automation Identification: Look for opportunities to automate routine aspects of your work, freeing up time and mental energy for higher-value activities.

Value Focus: Concentrate effort on activities that create the most value for employers, clients, or customers rather than trying to optimize all activities equally.

Energy Management: Recognize that human energy and attention are limited resources that should be allocated strategically rather than spread across all available tasks.

Rachel Kim, the marketing professional whose creative work was partially automated, found success through strategic laziness. "I stopped trying to write first drafts myself and started using AI to generate initial content that I could then edit and refine," she said.

> "This freed up mental energy for the strategic thinking - understanding client needs, developing campaign concepts, and figuring out how to connect with specific audiences. I was doing less busy work but creating more value."

Continuous Adaptation Without Constant Learning

One of the most exhausting aspects of working in a technology-influenced environment is the feeling that you need to constantly learn new tools, platforms, and systems. This creates what researchers call "learning fatigue" - the mental exhaustion that comes from continuous skill acquisition demands. Effective AI adaptation requires finding ways to remain current and effective without falling into the trap of constant learning. This involves developing meta-strategies that allow you to adapt to new AI tools without needing to master each one individually.

Pattern Recognition: Learn to identify common patterns across different AI tools and applications. Most AI systems share similar underlying approaches to user interaction, even if their specific interfaces differ.

Core Principle Focus: Instead of memorizing specific procedures for particular AI tools, focus on understanding the core principles of effective human-AI collaboration that apply across different systems.

Just-in-Time Learning: Develop skills in rapidly acquiring the specific knowledge you need for immediate tasks rather than trying to comprehensively understand every AI tool you might someday encounter.

Learning Community Participation: Join or create communities of people facing similar AI adaptation challenges. These communities can provide mutual support, share discoveries about effective strategies, and reduce the individual burden of staying current with technological developments.

Measuring Success in Human Terms

As work becomes increasingly AI-integrated, it's important to develop metrics for success that emphasize human value creation rather than trying to compete with AI on machine terms. This means redefining productivity, effectiveness, and professional growth in ways that highlight uniquely human contributions.

Relationship Quality: Measure success partly based on the strength and depth of professional relationships you build and maintain.

Problem-Solving Impact: Focus on the significance and complexity of problems you solve rather than the speed at which you process routine tasks.

Creative Contributions: Evaluate your work based on originality, innovation, and creative value rather than volume or consistency metrics where AI might excel.

Human Development: Consider your success in mentoring, teaching, and developing other people as a key metric of professional value.

Judgment Quality: Assess your contribution based on the quality of decisions you make in ambiguous or complex situations where AI guidance is limited.

Marcus Chen, the financial analyst we met at the beginning of Chapter 4, eventually found his path by redefining success in human terms. "I stopped trying to compete with AI on data processing speed and started focusing on strategic analysis and client communication," he said. "My success metrics changed from how many reports I could generate to how well I could help clients understand what the data meant for their specific situations. It was actually more fulfilling work, even though it took me a while to realize it was also more valuable."

Building Sustainable Adaptation Practices

The key to long-term success in an AI-integrated world is developing sustainable practices that allow continuous adaptation without burnout. This requires building systems and habits that support ongoing effectiveness without demanding constant major changes in approach or capabilities.

Regular Review Processes: Establish quarterly or semi-annual reviews of how AI is affecting your work, what new opportunities are emerging, and what adjustments might be beneficial.

Experimentation Protocols: Develop low-risk ways to test new approaches to human-AI collaboration, allowing you to learn and adapt without jeopardizing current effectiveness.

Network Maintenance: Invest regularly in maintaining and expanding professional relationships that can provide information, opportunities, and support during transitions.

Skill Portfolio Management: Maintain a diverse portfolio of capabilities that includes both AI-enhanced skills and AI-resistant human capabilities.

Stress Management Systems: Develop sustainable approaches to managing the stress and uncertainty that come with ongoing technological change.

The goal isn't to become an AI expert or to resist AI integration, but to find sustainable ways to remain valuable, effective, and fulfilled while working alongside increasingly sophisticated AI systems. This requires focusing on distinctly human capabilities while learning to leverage AI tools effectively - a balance that can be achieved without deep technical knowledge but does require strategic thinking about how to position yourself in an AI-enhanced world.

In our next chapter, we'll explore the darker aspects of this AI-integrated society - the potential risks, unintended consequences, and challenges that come with widespread AI adoption. Understanding these downsides is crucial for making informed decisions about how to navigate the AI revolution while protecting your interests and values.

6

THE DARK SIDE OF AI SOCIETY

"The price of progress is trouble, and I must be making a lot of progress." – Charles Kettering

When Rebecca Foster first noticed the changes in her teenage daughter Emma's behavior, she attributed them to typical adolescent moodiness. Emma had become withdrawn, spending hours in her room, seemingly absorbed in her phone or computer. Her grades hadn't suffered, but something felt different - a flatness in her interactions, a lack of engagement with family conversations, an almost robotic quality to her responses.

The revelation came during a family dinner when Emma casually mentioned that her AI tutoring app had helped her write an essay about climate change. "It's so much better than my original draft," Emma said, showing Rebecca a polished, sophisticated piece of writing that bore little resemblance to her daughter's usual voice.

> **"The AI understood exactly what I wanted to say and helped me say it better."**

Rebecca felt a chill of recognition. As a high school English teacher for fifteen years, she had been struggling with similar issues at work - students submitting work that seemed too polished, too perfect, lacking the characteristic mistakes and quirks that mark authentic human expression. But seeing it in her own home, watching her daughter's face light up as she described how AI had "improved" her thoughts, Rebecca realized they were all witnessing something unprecedented: a generation learning to outsource not just their work, but their thinking itself.

> **"I started wondering if we were raising a generation of intellectual dependents," Rebecca told me months later. "These kids are growing up believing that AI thoughts are better than human thoughts, that AI writing is better than human writing. They're not just using AI as a tool - they're using it as a crutch, and eventually, they might not be able to walk without it."**

Rebecca's concerns point to one of the most troubling aspects of our rapid AI integration: while we've focused heavily on the economic disruption and adaptation challenges, we've paid less attention to the deeper social, psychological, and cultural costs of creating an AI-dependent society. These costs aren't theoretical future problems - they're already manifesting in ways that should concern anyone who cares about human autonomy, creativity, and the preservation of essential human capabilities.

The Atrophy of Human Skills

Perhaps the most immediate and concerning consequence of widespread AI adoption is the systematic atrophy of human capabilities that people stop exercising because AI can perform them more efficiently. This phenomenon, which researchers call "automation-induced skill degradation," affects everything from basic arithmetic and writing skills to more complex abilities like critical thinking and creative problem-solving.

Dr. Matthew Crawford's research on attention and skill development reveals that capabilities require regular practice to maintain. "Skills are not like files stored on a computer," he explains. "They're more like muscles that weaken without use. When we consistently delegate thinking tasks to AI systems, we risk losing the very capabilities that make us distinctly human." The effects are already visible across multiple domains:

Mathematical Reasoning: Students who rely heavily on AI for mathematical problem-solving show decreased ability to work through problems manually. While AI can solve complex equations instantly, students are losing the mathematical intuition that comes from working through problems step by step.

Writing and Communication: Teachers report that students who use AI writing assistants extensively struggle more with original expression, logical argument development, and finding their authentic voice. The AI provides polished prose, but students miss the cognitive development that comes from wrestling with words and ideas.

Research and Information Evaluation: As AI provides instant answers to questions, people are exercising less discernment about information quality, source reliability, and logical reasoning. The skill of evaluating information critically atrophies when AI pre-processes and summarizes complex topics.

Creative Problem-Solving: When AI provides quick solutions to problems, people have fewer opportunities to develop the tolerance for ambiguity, persistence through difficulty, and creative insight that come from working through challenges independently.

Dr. Sarah Chen, who studies cognitive development at Johns Hopkins, has documented what she calls "intellectual learned helplessness" among heavy AI users. "We're seeing people who become anxious and uncertain when AI tools aren't available, even for tasks they previously handled with confidence," she reports. "It's as if they've forgotten that they're capable of independent thought and problem-solving." The concern isn't that people use AI tools - it's that they're beginning to rely on them for tasks that are essential for maintaining cognitive flexibility, creativity, and intellectual independence. When AI becomes a substitute for human thinking rather than an enhancement of it, the long-term consequences could be profound.

The Erosion of Human Agency

Beyond skill atrophy lies a more fundamental concern: the gradual erosion of human agency as AI systems make more decisions for us, often without our conscious awareness. This erosion happens subtly, through the accumulation of small conveniences that individually seem beneficial but collectively reduce our sense of control and self-determination.

Decision Outsourcing: AI recommendation systems increasingly influence what we watch, read, buy, and even whom we meet. While these recommendations are often helpful, they also narrow the range of choices we actively consider, potentially reducing our exposure to new ideas, experiences, and perspectives that might challenge or expand our thinking.

Preference Prediction: AI systems that predict our preferences based on past behavior can create feedback loops that reinforce existing patterns while discouraging exploration and growth. If Netflix always recommends movies similar to ones we've watched before, we might miss genres that could expand our interests.

Automated Routine Management: As AI takes over more routine decision-making - from scheduling appointments to managing finances to planning routes - people report feeling less connected to the details of their own lives. This convenience comes at the cost of the micro-decisions and small acts of self-determination that contribute to a sense of personal agency.

Algorithmic Authority: Perhaps most concerning is the growing tendency to defer to AI recommendations even in areas where human judgment should prevail. Studies show that people often accept AI suggestions about complex personal decisions—career moves, relationship advice, health choices - even when those suggestions are based on limited or inappropriate data.

James Morrison, a 34-year-old marketing manager, described his growing unease with AI dependency: "I realized I was asking AI for advice about everything - what to cook for dinner, how to handle a conflict with my boss, whether I should break up with my girlfriend. It was supposed to make my life easier, but I started feeling like I wasn't living my own life anymore. I was just following AI suggestions." This erosion of agency is particularly troubling because it often feels empowering in the short term. AI systems are designed to provide helpful suggestions and make life more convenient. The problem emerges gradually, as people realize they've lost touch with their own preferences, judgment, and decision-making processes.

The Social Fragmentation Effect

AI integration is also contributing to increased social fragmentation in ways that may undermine community cohesion and shared understanding. As AI systems become better at providing personalized experiences, they can inadvertently isolate people in customized information bubbles that reduce common ground for social connection and democratic discourse.

Personalization Paradox: While AI personalization makes individual experiences more relevant and efficient, it can reduce the shared cultural references and common experiences that bind communities together. When everyone receives different news, entertainment, and information based on AI algorithms, the foundation for shared social understanding erodes.

Empathy Deficit: As AI handles more interpersonal communication - from customer service to social media interactions - people have fewer opportunities to practice reading human emotions, understanding different perspectives, and developing the empathy that comes from genuine human interaction.

Relationship Substitution: The rise of AI companions, chatbots, and virtual assistants that can simulate human conversation creates the risk that some people might prefer AI relationships to human ones. While these AI interactions can be comforting and convenient, they lack the unpredictability, challenge, and growth opportunities that come from real human relationships.

Community Skill Loss: As AI systems handle more coordination, communication, and organization tasks, communities may lose the collaborative skills needed for collective action, democratic participation, and social problem-solving.

Dr. Robert Putnam's research on social capital suggests that these trends could accelerate the social disconnection that has already been documented in many communities. "AI might solve individual problems more efficiently than human cooperation," he notes, "but it can't provide the social bonds, mutual support, and collective efficacy that come from people working together to address shared challenges."

The Authenticity Crisis

One of the most philosophically troubling aspects of AI integration is what I call the "authenticity crisis" - the growing difficulty of distinguishing between human-generated and AI-generated content, thoughts, and expressions. This crisis affects not just external verification of authenticity but our internal sense of what constitutes genuine human expression.

Creative Devaluation: As AI systems produce art, music, writing, and other creative works that are indistinguishable from human creations, questions arise about the value and meaning of human creativity. If an AI can write a poem that moves people to tears, what makes human poetry special?

Expression Uncertainty: People who use AI writing assistants report uncertainty about which ideas are truly their own. When AI helps refine thoughts, improve arguments, and polish expression, the line between human and artificial contribution becomes blurred.

Performance Pressure: The ability of AI to produce polished, professional-quality work creates pressure for humans to match that level of perfection in their own output. This can lead to increased anxiety, reduced willingness to share imperfect work, and decreased tolerance for the messiness that characterizes authentic human expression.

Identity Confusion: Young people who grow up using AI tools for creative and intellectual tasks may struggle with questions of authentic self-expression. If AI helps with writing, thinking, and problem-solving from an early age, how do people develop a sense of their own intellectual identity?

Maria Santos, now working as an escalation specialist after her customer service job was automated, described her daughter's struggle with this issue: "She asked me how she could know if her thoughts were really hers when she used AI to help with everything from homework to creative writing. It broke my heart because I realized she was growing up without confidence in her own mind."

The Inequality Amplification Problem

While AI has the potential to democratize access to information and capabilities, in practice it often amplifies existing inequalities rather than reducing them. This happens because the benefits of AI integration are not distributed equally across different socioeconomic groups, educational levels, or geographic regions.

Digital Divide Expansion: Access to advanced AI tools often requires reliable internet, modern devices, and digital literacy skills that are not equally distributed. This means that the people who could benefit most from AI assistance may be the least likely to have access to it.

Educational Stratification: Students in well-funded schools with access to cutting-edge AI tools gain significant advantages over those in under-resourced educational environments. This could accelerate educational inequality rather than reducing it.

Employment Polarization: AI tends to eliminate middle-skill jobs while creating opportunities that require either very high skills (AI development, complex analysis) or very low skills (tasks requiring physical presence and human interaction). This polarization can hollow out the middle class.

Resource Concentration: Companies and individuals with the resources to acquire and effectively use AI tools gain competitive advantages that can compound over time, potentially increasing wealth and power concentration.

Cultural Bias Reinforcement: AI systems trained on data that reflects historical inequalities and cultural biases can perpetuate and amplify those biases in their recommendations, decisions, and outputs.

Dr. Safiya Noble's research on algorithmic bias reveals how these inequalities can become embedded in AI systems in ways that are difficult to detect and correct. "AI systems don't just reflect the biases in their training data," she explains. "They can amplify and systematize those biases, making discrimination more efficient and harder to challenge."

The Dependency Trap

Perhaps the most concerning long-term risk of AI integration is what economists call the "dependency trap"—a situation where individuals, organizations, or entire societies become so reliant on AI systems that they lose the ability to function effectively without them, even as those systems may not be serving their best interests.

Vendor Lock-in: As people and organizations integrate AI tools deeply into their workflows and decision-making processes, switching to alternatives becomes increasingly difficult and expensive. This can create situations where users remain dependent on AI systems even when those systems become less effective or more expensive.

Skill Replacement: When AI systems replace human capabilities rather than augmenting them, users can lose the ability to perform essential tasks independently. This creates vulnerability if AI systems become unavailable, unreliable, or compromised.

Cognitive Outsourcing: The convenience of having AI handle complex thinking tasks can lead to a form of cognitive outsourcing where people lose confidence in their own judgment and reasoning abilities.

System Brittleness: Heavy dependence on AI systems can make individuals and organizations more vulnerable to system failures, cyber attacks, or changes in AI availability or functionality.

Economic Extraction: As dependence on AI systems increases, the companies controlling those systems gain more power to extract value from users through higher prices, reduced functionality, or data exploitation.

The dependency trap is particularly insidious because it develops gradually and often feels empowering in the early stages. AI systems genuinely do make many tasks easier and more efficient. The problem emerges when that efficiency comes at the cost of human agency, capability, and independence.

The Surveillance Society

AI integration often comes with increased monitoring, data collection, and behavioral tracking that can create unprecedented levels of surveillance in daily life. While this surveillance is usually presented as beneficial - providing personalized services, improving security, optimizing experiences - it also has profound implications for privacy, autonomy, and human freedom.

Behavioral Monitoring: AI systems that optimize performance often require detailed tracking of user behavior, creating comprehensive profiles of individual activities, preferences, and patterns.

Predictive Control: AI systems that predict behavior can be used not just to provide services but to influence and control behavior in ways that users may not recognize or consent to.

Social Credit Systems: The combination of AI analysis and behavioral tracking can create systems where people's access to services, opportunities, or resources depends on algorithmic assessments of their behavior and choices.

Workplace Monitoring: AI-enabled workplace monitoring can track employee productivity, communication patterns, and even emotional states in ways that reduce privacy and increase stress.

Democratic Erosion: Comprehensive surveillance systems, even when initially designed for beneficial purposes, can be repurposed for political control and the suppression of dissent.

The surveillance concerns are not just about privacy - they're about the kind of society we're creating and whether it will preserve the conditions necessary for human flourishing, creativity, and democratic participation.

The Meaning Crisis

As AI systems become capable of performing more tasks that humans have traditionally found meaningful - creating art, solving complex problems, providing care and companionship - questions arise about what gives human life purpose and significance.

Work-Based Identity Erosion: For people whose sense of purpose comes primarily from their professional contributions, AI automation can create existential as well as economic challenges.

Creative Devaluation: When AI can produce creative works that are indistinguishable from human creations, some people question the value and significance of human creativity.

Intellectual Redundancy: The feeling that AI can think faster, more accurately, and more comprehensively than humans can lead to a sense of intellectual inadequacy and meaninglessness.

Purpose Confusion: As AI handles more tasks that previously provided structure, challenge, and satisfaction in human life, people may struggle to identify sources of meaning and purpose.

Existential Questioning: The rapid advancement of AI capabilities raises fundamental questions about what makes humans special, valuable, or necessary in an increasingly automated world.

These meaning-related challenges are not just individual psychological issues - they can have broader social consequences including increased depression, social withdrawal, political extremism, and destructive behaviors.

Protecting Against the Dark Side

Understanding these risks doesn't require rejecting AI technology entirely, but it does suggest the need for intentional strategies to protect against the negative consequences of AI integration while still benefiting from its positive potential.

Capability Preservation: Deliberately practice skills that AI can perform to maintain human capabilities even when AI assistance is available. This might include manual calculation, handwriting, critical reading, and independent problem-solving.

Agency Maintenance: Regularly make decisions without AI assistance, even for tasks where AI could be helpful. Maintain the habit of independent judgment and choice-making.

Community Investment: Prioritize human relationships and community participation that can provide meaning, support, and shared experience beyond what AI systems can offer.

Digital Boundaries: Establish clear limits on AI use, create AI-free spaces and times, and maintain activities that require purely human engagement.

Critical Evaluation: Develop skills in evaluating AI outputs, questioning AI recommendations, and maintaining healthy skepticism about algorithmic authority.

Values Clarification: Regularly reflect on personal values, goals, and sources of meaning that exist independently of technological capabilities.

Privacy Protection: Take active steps to limit data collection, maintain privacy, and resist unnecessary surveillance, even when it comes with convenience benefits.

The dark side of AI society is not inevitable - it emerges from choices about how we integrate AI systems into our lives and communities. By understanding these risks and taking deliberate action to mitigate them, we can work toward a future that harnesses AI's benefits while preserving essential human capabilities, relationships, and freedoms. In our next chapter, we'll explore the brighter possibilities - the uniquely human skills and capabilities that AI cannot replace and how strengthening these can provide not just protection against AI's negative effects but pathways to enhanced human flourishing in an AI-integrated world.

7

THE BRIGHT SIDE: HUMAN SKILLS AI CAN'T REPLACE

"The best way to find out if you can trust somebody is to trust them." – Ernest Hemingway

When 78-year-old Eleanor Martinez was diagnosed with early-stage dementia, her family faced a difficult decision about her care. The assisted living facility they toured showcased impressive AI-powered systems: medication dispensers that ensured perfect dosing schedules, fall-detection sensors that could summon help within seconds, and cognitive assessment tools that tracked mental decline with scientific precision. The technology was undeniably advanced and potentially life-saving. But what convinced Eleanor's daughter Carmen to choose a different facility wasn't the absence of technology - it was the presence of something irreplaceable.

During their visit, Carmen watched as a nurse named Patricia sat with Eleanor for twenty minutes, not checking vitals or administering medication, but simply listening as Eleanor told a rambling story about her late husband's garden. Patricia asked gentle questions, laughed at appropriate moments, and somehow helped Eleanor feel heard and valued despite her confusion.

"The AI systems could monitor my mother's body perfectly," Carmen told me later. "But Patricia saw my mother's soul. She understood that beneath the cognitive decline was a person who needed dignity, connection, and the feeling that her stories still mattered. No algorithm can provide that kind of human recognition."

Patricia's interaction with Eleanor illustrates one of the most encouraging aspects of our AI-integrated future: there are fundamentally human capabilities that artificial intelligence cannot replicate, no matter how sophisticated it becomes. While previous chapters have explored the challenges and risks of AI adoption, this chapter focuses on the bright side - the irreplaceable human skills that become more valuable, not less, as AI handles routine cognitive tasks.

Understanding these capabilities isn't just about career security or economic survival. It's about recognizing what makes human life meaningful and how those qualities can flourish even in - perhaps especially in - a world filled with artificial intelligence.

The Irreplaceable Core of Human Experience

At the heart of irreplaceable human capabilities lies something that AI researchers call the "hard problem of consciousness" - the subjective, experiential quality of being human that cannot be replicated by even the most sophisticated information processing systems. This isn't just philosophical speculation; it has practical implications for what humans can offer that AI systems cannot.

Subjective Experience: Humans don't just process information about emotions, relationships, or aesthetic experiences - we actually feel them. This subjective experience creates insights, judgments, and responses that emerge from lived experience rather than data analysis.

Embodied Understanding: Human knowledge is shaped by having a physical body that moves through the world, experiences pleasure and pain, and interacts with other embodied beings. This creates forms of understanding that transcend purely intellectual knowledge.

Mortality Awareness: The knowledge that life is finite and precious gives human perspectives a weight and urgency that AI systems, which exist in a kind of eternal present, cannot replicate.

Intentionality and Purpose: Humans act not just based on optimization algorithms but from deep purposes, values, and meanings that emerge from the complex interplay of consciousness, experience, and choice. These fundamental aspects of human experience create capabilities that manifest in practical, valuable ways across numerous domains of work and life.

The Power of Authentic Empathy

Perhaps the most immediately valuable irreplaceable human skill is authentic empathy - the ability to genuinely understand and share the feelings of another person based on your own emotional experience. While AI systems can simulate empathy and even provide helpful responses based on emotional data, they cannot actually feel what others are feeling. This authentic empathy manifests in numerous practical applications:

Healthcare and Healing: Medical AI can diagnose conditions and recommend treatments with increasing accuracy, but human healthcare providers offer something fundamentally different—the comfort and healing that comes from being understood by another conscious being who has experienced vulnerability, fear, and hope.

Dr. Michael Chen, an oncologist who works alongside AI diagnostic tools, describes the irreplaceable human element: "The AI can tell me exactly what treatment protocol has the highest success rate for a specific type of cancer. But when I sit with patients after delivering a difficult diagnosis, I'm drawing on my own experiences with loss, my own fears about mortality, my own understanding of what it means to face uncertainty. That shared humanity is healing in ways that go beyond medical intervention."

Education and Development: AI tutoring systems can personalize learning and provide immediate feedback, but human teachers offer something AI cannot - the experience of being a learner themselves, of struggling with difficult concepts, of feeling the joy of understanding. This creates a different kind of learning relationship.

Sarah Wilson, a high school mathematics teacher who now works alongside AI teaching assistants, found that her role evolved rather than diminished: "The AI handles a lot of the routine explanation and practice problems, but I'm the one who can see when a student is frustrated not just with math but with feeling stupid. I remember what it was like to struggle with algebra, and I can offer encouragement and perspective that comes from having been there myself."

Counseling and Support: While AI chatbots can provide 24/7 availability and consistent responses based on therapeutic techniques, human counselors offer the irreplaceable experience of being genuinely understood by someone who has navigated their own emotional challenges and growth.

Customer Relationships: In business contexts, authentic empathy allows humans to understand not just what customers are asking for, but what they're actually worried about, hoping for, or trying to achieve. This deeper understanding creates trust and loyalty that transactional AI interactions cannot match.

Creative and Aesthetic Intelligence

Human creativity operates fundamentally differently from AI-generated content, even when the outputs appear similar. Human creativity emerges from the intersection of conscious experience, emotional depth, cultural context, and the meaning-making that comes from being a mortal, embodied being navigating an uncertain world.

Lived Experience as Creative Source: Human artists, writers, and creators draw inspiration from their own experiences of love, loss, struggle, triumph, and wonder. These experiences provide not just material for creative work but emotional authenticity that resonates with others who share similar experiences.

Cultural Translation and Bridge-Building: Humans who have lived within specific cultural contexts can create work that authentically represents those experiences while making them accessible to others. This cultural translation requires not just knowledge about a culture but lived experience within it.

Aesthetic Judgment in Context: While AI can analyze patterns in what people have liked previously, humans can make aesthetic judgments based on intuition, emotional response, and cultural understanding that goes beyond pattern matching.

Meaningful Innovation: Human creativity often aims to express something meaningful about the human condition, challenge social assumptions, or provide comfort during difficult times. This purpose-driven creativity serves functions beyond entertainment or aesthetic pleasure.

Marcus Rivera, a graphic designer who initially feared AI would eliminate his profession, discovered that his role evolved toward more strategic and meaningful work: "AI can generate dozens of logo concepts in minutes, and some of them are technically excellent. But I'm the one who understands what the business owner is really trying to communicate about their values, their community, their hopes for their company. I help them figure out which direction feels authentic to their story, not just which design tests well with focus groups."

Moral and Ethical Reasoning

Perhaps no area highlights the irreplaceable nature of human intelligence more clearly than moral and ethical reasoning. While AI systems can be programmed with ethical guidelines and can identify potential ethical issues, they cannot engage in the kind of moral reasoning that emerges from conscious experience of right and wrong, the capacity to suffer, and the weight of moral responsibility.

Contextual Moral Judgment: Human moral reasoning takes into account not just rules and principles but the complex contextual factors that affect ethical decisions - the relationships involved, the historical context, the unintended consequences, and the competing values at stake.

Moral Innovation: Humans can recognize when existing moral frameworks are insufficient and develop new approaches to ethical challenges. This moral innovation requires the kind of creative ethical thinking that emerges from grappling with genuine moral dilemmas.

Ethical Leadership: Leading others through complex ethical challenges requires not just knowledge of ethical principles but the moral authority that comes from having struggled with difficult decisions and taken responsibility for their consequences.

Value Integration: Humans can integrate multiple, sometimes competing values - justice and mercy, efficiency and fairness, individual rights and community good - in ways that reflect the complexity of real-world moral situations.

Jennifer Walsh, the financial advisor who adapted to AI-enhanced portfolio management, found that her most valuable contributions involved ethical guidance: "The robo-advisors can optimize portfolios for maximum return, but they can't help clients think through whether their investments align with their values. I've helped clients navigate questions about sustainable investing, about supporting businesses that match their beliefs, about balancing financial goals with ethical considerations. Those conversations require someone who has thought deeply about the relationship between money and values."

Relationship Building and Trust Creation

Trust - genuine, deep trust between human beings - remains one of the most uniquely human capabilities. While AI systems can be reliable and consistent, the kind of trust that enables deep relationships, effective teamwork, and community building requires vulnerability, reciprocity, and the mutual recognition of shared humanity.

Vulnerability and Authenticity: Building deep trust requires the ability to be vulnerable - to share doubts, fears, and imperfections. This vulnerability creates emotional bonds that are fundamentally different from the reliability that AI systems provide.

Mutual Recognition: Trust between humans involves recognizing and valuing each other as conscious, experiencing beings with their own hopes, fears, and dignity. This mutual recognition creates a foundation for cooperation that goes beyond transactional relationships.

Forgiveness and Second Chances: Human relationships can survive mistakes, misunderstandings, and even betrayals through forgiveness and the possibility of growth and change. This capacity for redemption and renewal is distinctly human.

Long-term Commitment: Humans can make and keep commitments that extend beyond immediate optimization, creating stable relationships and communities that persist through changing circumstances.

David Park, the journalist who adapted to AI-generated news content, found that his most valuable work involved relationship building: "The AI could generate accurate reports about city council meetings, but I was the one who built trust with community members over years of fair reporting. When something controversial happened, people would call me because they trusted me to understand the full context and report honestly. That trust took years to build and couldn't be replicated by any algorithm."

Wisdom and Life Integration

Wisdom - the integration of knowledge, experience, and judgment that comes from living through diverse life experiences - represents perhaps the most sophisticated irreplaceable human capability. Wisdom involves not just knowing facts or following rules, but understanding how to apply knowledge appropriately in complex, ambiguous, real-world situations.

Pattern Recognition Across Domains: Human wisdom integrates patterns from multiple life domains - professional experience, personal relationships, historical knowledge, and cultural understanding - in ways that create insights beyond what specialized AI systems can provide.

Temporal Perspective: Humans who have lived through multiple decades can offer perspectives on how situations, trends, and problems evolve over time. This long-term perspective is invaluable for decision-making in uncertain circumstances.

Paradox Integration: Wisdom often involves holding contradictory truths simultaneously and knowing when to apply different principles in different circumstances. This comfort with paradox and ambiguity is distinctly human.

Life Meaning and Purpose: Wise humans can help others navigate questions of meaning, purpose, and life direction that go beyond optimization problems and touch on the deepest questions of human existence.

Intergenerational Connection: The ability to understand and bridge different generations, translating insights from one age group to another, requires the kind of empathetic understanding that comes from human life experience.

Eleanor Martinez's nurse Patricia exemplified this kind of wisdom. She understood that caring for someone with dementia required not just medical knowledge but wisdom about human dignity, the importance of personal history, and the ways that respect and kindness can provide comfort even when cognitive function is declining.

Intuitive Problem Solving

Human intuition - the ability to understand or know something immediately, without conscious reasoning - represents a sophisticated form of intelligence that AI systems cannot replicate. This intuition emerges from the integration of conscious and unconscious processing, embodied experience, and pattern recognition that occurs below the threshold of awareness.

Rapid Pattern Recognition: Experienced humans can often recognize complex patterns and situations almost instantly, drawing on years of experience integrated at an unconscious level.

Emotional Intelligence: The ability to read subtle emotional cues, understand interpersonal dynamics, and navigate complex social situations requires intuitive understanding that goes beyond logical analysis.

Creative Leaps: Human creativity often involves intuitive leaps that connect seemingly unrelated ideas or concepts in novel ways. These creative insights emerge from the unconscious integration of diverse experiences and knowledge.

Situation Assessment: In complex, ambiguous situations, humans can often intuitively grasp the essential elements and appropriate responses faster than systematic analysis would allow.

Risk Evaluation: Human intuition about risk often incorporates subtle factors that are difficult to quantify but crucial for good decision-making.

Tom Richards, the radiologist working with AI diagnostic tools, found that his intuitive abilities became more valuable, not less: "The AI is excellent at identifying specific patterns it's been trained to recognize. But sometimes I'll look at an image and something just feels wrong, even though I can't immediately articulate what it is. That intuitive sense, developed over decades of looking at thousands of images, often leads to discoveries that the AI missed."

Cultural and Community Intelligence

Humans embedded in specific cultures, communities, and social contexts possess irreplaceable knowledge about the nuances, values, relationships, and dynamics that shape how things actually work in those environments. This cultural intelligence cannot be replicated by AI systems trained on general data.

Local Knowledge: Understanding how things actually work in specific communities - who the real decision-makers are, what the unspoken rules are, how to navigate local customs and relationships - requires lived experience within those communities.

Cultural Translation: The ability to help people from different backgrounds understand each other requires not just knowledge about different cultures but the empathetic understanding that comes from navigating cultural differences personally.

Community Leadership: Leading and organizing within communities requires understanding the history, relationships, conflicts, and aspirations that shape community dynamics. This understanding develops through participation and relationship-building over time.

Social Innovation: Creating new approaches to community problems requires understanding not just what the problems are but how the community's values, resources, and dynamics can be leveraged for positive change.

Conflict Resolution: Mediating disputes and building consensus within communities requires understanding the emotional, historical, and relational factors that contribute to conflict and the pathways to healing and cooperation.

The Integration Advantage

Perhaps most importantly, humans excel at integration - bringing together insights, experiences, and capabilities from multiple domains to address complex, multi-faceted challenges. While AI systems typically excel within specific domains, humans can synthesize across domains in ways that create novel solutions and approaches.

Cross-Domain Synthesis: Humans can apply insights from one field to challenges in completely different areas, creating innovative solutions that single-domain AI systems might miss.

Holistic Problem Solving: Complex real-world problems often require addressing technical, emotional, social, and ethical dimensions simultaneously. Humans can hold all of these dimensions in mind and develop integrated solutions.

Adaptive Strategy: Humans can adjust their approach based on changing circumstances, new information, and emergent challenges in ways that demonstrate cognitive flexibility beyond current AI capabilities.

Value Integration: When multiple values or priorities conflict, humans can navigate these tensions and find approaches that honor competing concerns in contextually appropriate ways.

Developing and Leveraging Irreplaceable Skills

Understanding what makes humans irreplaceable is only the beginning. The next step involves deliberately developing these capabilities and learning to leverage them effectively in an AI-integrated world.

Conscious Cultivation: Rather than taking these human capabilities for granted, consciously develop them through practice, reflection, and learning from others who demonstrate these skills effectively.

Integration Strategy: Look for ways to combine these uniquely human capabilities with AI tools to create value that neither humans nor AI could provide alone.

Market Positioning: Position yourself professionally in roles that require and reward these irreplaceable human capabilities rather than trying to compete with AI in areas where it excels.

Community Building: Create and participate in communities that value and nurture these human capabilities, providing contexts where they can flourish and be recognized.

Continuous Development: Treat these capabilities as skills that can be developed throughout life rather than fixed traits. Seek out experiences, relationships, and challenges that stretch and strengthen these uniquely human abilities.

The future doesn't belong to humans who can compete with AI systems at information processing, pattern recognition, or optimization. It belongs to humans who can do what AI cannot - love authentically, create meaningfully, judge wisely, and connect genuinely with other conscious beings navigating the profound mystery of existence. These irreplaceable capabilities aren't just economic assets or career insurance. They're the foundation of what makes life meaningful, relationships fulfilling, and communities thriving.

In a world increasingly filled with artificial intelligence, the most important intelligence of all may be the ancient, irreplaceable intelligence of being fully, consciously, courageously human. In our next chapter, we'll explore how to build a new identity that integrates these irreplaceable human capabilities with the practical realities of living and working alongside AI systems. The goal isn't to retreat from technology but to engage with it from a position of human strength and clarity about what we uniquely bring to the world.

8

BUILDING A NEW IDENTITY

"The curious paradox is that when I accept myself just as I am, then I can change." – Carl Rogers

For thirty-two years, Robert Chen had introduced himself the same way: "I'm an accountant." It wasn't just his job title - it was his identity, his source of pride, his answer to the question "Who are you?" He'd built his sense of self around precision, reliability, and the satisfaction of making numbers balance perfectly. His office walls displayed his CPA certification, awards for accuracy, and photos from company events where he'd been recognized for his meticulous work.

Then, in early 2024, his firm implemented an AI system that could process tax returns in minutes rather than hours, catch errors he might miss, and generate reports with flawless formatting. Initially, Robert tried to compete - working longer hours, double-checking the AI's work, finding small mistakes to prove his continued value. But gradually, he realized he was fighting a losing battle against a system that was not only faster and more accurate but improving every month.

The breaking point came during a client meeting when Mrs. Patterson, a longtime client, thanked him for his years of service but mentioned that she was switching to an online AI tax service. "It's just so convenient," she said apologetically, "and honestly, I get the same results for a fraction of the cost." Walking home that evening, Robert experienced what he later described as an identity crisis. "If I wasn't an accountant," he told me, "then who was I? Everything I'd built my sense of self around - my expertise, my reliability, my professional identity - suddenly felt obsolete. It wasn't just about losing a job; it was about losing myself."

Robert's experience illustrates one of the most profound challenges of the AI revolution: the need to reconstruct our identities for a world where traditional sources of identity - career, expertise, economic function - may no longer provide the stable foundation they once did. This reconstruction isn't just a practical career challenge; it's a deep psychological and spiritual work that involves rediscovering who we are beyond our economic roles.

The Crisis of Professional Identity

For generations, the answer to "What do you do?" has served as shorthand for "Who are you?" Our professional identities have provided not just income but also social status, personal meaning, community connection, and a framework for understanding our place in the world. When AI disrupts these professional identities, it creates an existential challenge that goes far beyond unemployment. Dr. William Bridges, a researcher on life transitions, identifies three stages in any major identity shift: endings, the neutral zone, and new beginnings. In the context of AI disruption, these stages play out in specific ways:

The Ending: This involves letting go of identities that no longer serve us. For many people, this means accepting that being "the Excel expert" or "the person who knows all the procedures" may no longer be valuable. The ending stage often involves grief, anger, and resistance as people mourn the loss of identities that provided security and meaning.

The Neutral Zone: This uncomfortable middle period involves living without a clear professional identity while exploring new possibilities. People in this stage often report feeling lost, uncertain, and anxious about the future. However, this discomfort is also where creativity and new possibilities emerge.

New Beginnings: This final stage involves constructing new identities based on a deeper understanding of personal values, capabilities, and purposes that extend beyond narrow professional roles.

The key insight from Bridges' research is that trying to skip directly from an old identity to a new one without going through the neutral zone usually fails. The discomfort of not knowing who we are professionally is actually a necessary part of discovering who we might become.

Beyond Professional Labels

One of the most liberating realizations in building a new identity is that human worth and identity extend far beyond professional labels. While career disruption can feel catastrophic, it also creates opportunities to develop more multifaceted, resilient identities that aren't dependent on any single role or economic function.

Core Values Identification: Rather than defining yourself by what you do, identify what you value most deeply. Are you someone who values justice, creativity, connection, learning, service, beauty, truth, freedom? These values can provide stable identity anchors that persist across changing circumstances.

Character Strengths Recognition: Research by psychologist Martin Seligman identifies 24 character strengths that appear across cultures and throughout history. These include courage, kindness, curiosity, perseverance, fairness, gratitude, and leadership. Unlike job skills, these character strengths represent enduring aspects of who you are.

Relationship-Based Identity: Consider how your identity is shaped by your relationships - as a parent, friend, mentor, community member, or partner. These relational identities often provide more meaning and stability than professional roles.

Life Experience Integration: Your unique combination of experiences, challenges overcome, lessons learned, and wisdom gained constitutes an irreplaceable identity foundation that no AI can replicate.

Purpose Beyond Paycheck: Identify the underlying purpose or contribution you want to make in the world, separate from how you currently make money. This purpose can be fulfilled through various means and isn't dependent on specific job titles or industries.

Maria Santos, the former customer service representative we've followed throughout this book, described her identity transformation: "I realized I'd always defined myself as 'someone who helps people solve problems,' but I'd confused that with 'someone who works in customer service.' Once I separated my core identity from my job title, I could see lots of ways to help people solve problems - through volunteer work, community organizing, even helping my neighbors navigate technology. The 'what' could change, but the 'who' remained constant."

The Portfolio Identity Model

In an AI-dominated world, the most resilient identities may be what researchers call "portfolio identities" - multifaceted self-concepts that draw strength and meaning from multiple sources rather than relying on a single professional role.

Multiple Competencies: Instead of being "an accountant" or "a teacher," develop an identity that encompasses multiple competencies. You might be "someone who combines analytical thinking, communication skills, and community leadership" or "a person who integrates creative problem-solving with technical understanding and mentoring abilities."

Diverse Contribution Streams: Create identity around multiple ways you contribute value - through paid work, volunteer activities, family roles, creative pursuits, and community involvement. This diversification provides both meaning and resilience when any single stream is disrupted.

Adaptive Capacity: Include "someone who adapts well to change" or "a person who learns quickly" as core identity elements. This meta-capacity helps you navigate future disruptions while maintaining a stable sense of self.

Values-Based Integration: Use your core values as the thread that connects diverse activities and roles. If you value justice, you might pursue it through legal work, community organizing, volunteer advocacy, and raising socially conscious children.

Growth Orientation: Build identity around being someone who is always growing, learning, and evolving rather than someone who has achieved a fixed set of accomplishments.

Jennifer Walsh, the financial advisor who learned to work alongside robo-advisors, developed a portfolio identity that served her well: "I stopped thinking of myself as 'a financial advisor' and started seeing myself as 'someone who helps people make wise decisions about money and life.' That identity could be fulfilled through traditional financial planning, but also through teaching financial literacy classes, writing about personal finance, mentoring young professionals, or even helping family members think through major financial decisions."

The Integration Challenge

One of the most complex aspects of building a new identity in an AI world involves integrating human capabilities with AI tools in ways that create coherent, authentic self-concepts. This isn't about becoming a "human-AI hybrid" but about maintaining human identity while effectively leveraging AI assistance.

Tool vs. Identity Boundaries: Learn to see AI as sophisticated tools that enhance your capabilities rather than extensions of your identity. Just as a chef's identity isn't defined by their knives but by their creativity, skill, and ability to nourish others, your identity shouldn't be defined by the AI tools you use but by the uniquely human value you create.

Amplification vs. Replacement: Frame AI as amplifying your human capabilities rather than replacing them. If you're someone who values helping others learn, AI tutoring tools might amplify your ability to provide personalized education rather than threatening your identity as an educator.

Collaboration Mindset: Develop an identity as someone who collaborates effectively with both humans and AI systems. This positions you as a bridge between human needs and technological capabilities rather than as competition for AI systems.

Meta-Skill Integration: Include proficiency in directing, evaluating, and integrating AI outputs as part of your identity. This frames AI fluency as an extension of your human judgment rather than a separate technical skill.

Ethical Foundation: Maintain strong ethical principles about how and when to use AI tools as part of your identity. This ethical foundation helps preserve human agency and values while engaging with AI systems.

Narrative Identity Development

Psychologists have discovered that people who navigate major life transitions most successfully often do so by reconstructing their "narrative identity" - the internal story they tell about who they are, where they've been, and where they're going. In the context of AI disruption, developing a new narrative identity becomes crucial for psychological well-being and effective adaptation.

Continuity Recognition: Identify the threads of continuity that connect your past, present, and future. What core aspects of who you are have remained constant even as circumstances changed? These continuities provide stability during identity reconstruction.

Growth Story Creation: Frame career disruption and adaptation as part of a growth story rather than a decline narrative. Instead of "I lost my job to AI," the story might be "I discovered new ways to apply my talents" or "I learned to focus on what I do best."

Challenge Integration: Include past challenges and how you overcame them as evidence of your resilience and adaptability. This builds confidence for handling current AI-related disruptions.

Purpose Articulation: Develop a clear story about your purpose or mission in life that extends beyond any particular job or role. This purpose provides direction and meaning during periods of professional uncertainty.

Future Visioning: Create compelling stories about possible futures that excite and motivate you. These don't need to be detailed career plans but rather inspiring visions of how you might contribute to the world.

Robert Chen, the accountant whose story opened this chapter, eventually developed a new narrative identity that served him well: "I realized that what I'd always loved about accounting wasn't really the numbers - it was helping small business owners understand their financial situation so they could make better decisions. I started seeing myself as a 'financial translator' who helps non-financial people understand what their numbers mean. AI can process the data, but I help people understand what the data means for their dreams, their risks, and their decisions."

Identity Experiments and Safe Exploration

Building a new identity doesn't happen overnight, and it's often helpful to approach it as a series of experiments rather than trying to make one dramatic change. Safe exploration allows you to test new identity elements without abandoning everything familiar.

Volunteer Exploration: Volunteer activities provide low-risk ways to explore new identity elements. If you're curious about whether "educator" might be part of your identity, volunteer to teach or tutor in your area of expertise.

Side Project Development: Develop small side projects that allow you to explore different aspects of your identity. These might include creative pursuits, community initiatives, or small business ventures.

Role Expansion: Look for ways to expand your current role to include identity elements you want to develop. If you want to integrate "mentor" into your identity, seek opportunities to guide newer employees or colleagues.

Community Involvement: Participate in communities aligned with identity elements you want to develop. If "community leader" appeals to you, start attending local meetings or joining community organizations.

Skill Development: Learn new skills not just for career advancement but for identity exploration. Taking a writing class might reveal a "storyteller" identity; learning facilitation skills might uncover a "group leader" identity.

Identity Interviews: Talk to people who embody identity elements that interest you. Learn about their paths, challenges, and satisfactions to help you evaluate whether these identity elements might fit you.

The key is to approach these experiments with curiosity rather than pressure. The goal isn't to find the perfect new identity quickly but to gradually discover what resonates with your deepest values and authentic self.

The Role of Community in Identity Formation

Identity isn't formed in isolation - it emerges through interaction with others and is validated through community recognition and response. Building a new identity in an AI-disrupted world often requires finding or creating communities that support and recognize the identity elements you're developing.

Professional Communities: Join professional associations, online groups, or local meetups related to new identity elements you're exploring. If you're developing an identity as a "learning facilitator," connect with adult education professionals or training specialists.

Values-Based Communities: Find communities organized around your core values rather than your job title. If you value environmental sustainability, connect with local environmental groups; if you value social justice, engage with advocacy organizations.

Peer Support Groups: Connect with others going through similar identity transitions. These might be formal support groups for career changers or informal networks of people adapting to AI disruption in your industry.

Mentoring Relationships: Both being mentored and mentoring others can support identity development. Learning from others who have made similar transitions provides guidance, while mentoring others helps solidify your own growth and new identity elements.

Creative Communities: If creativity is becoming part of your identity, engage with local arts communities, writing groups, maker spaces, or other creative collectives.

Service Communities: If service to others is a core identity element, find volunteer organizations or service-oriented groups where this aspect of your identity can be expressed and validated.

Managing Identity Transition Challenges

Building a new identity while navigating AI disruption creates specific challenges that require deliberate management strategies.

Imposter Syndrome: As you explore new identity elements, you may feel like a fraud or worry that others will discover you don't "really" belong in new roles or communities. Remember that everyone feels uncertain when developing new aspects of themselves - competence follows engagement, not the reverse.

Identity Diffusion: With multiple identity possibilities, you might feel scattered or unclear about who you're becoming. This is normal during transition periods. Focus on underlying values and purposes that can integrate diverse identity elements.

Social Pressure: Friends, family, or former colleagues might pressure you to return to familiar identity patterns or express skepticism about changes you're making. Prepare responses that acknowledge their concerns while maintaining commitment to your growth.

Financial Anxiety: Identity reconstruction often happens alongside financial uncertainty, which can make it difficult to make decisions based on authentic identity rather than immediate economic needs. Where possible, maintain some financial stability while exploring new identity possibilities.

Timeline Pressure: You might feel pressured to complete your identity reconstruction quickly, especially if career changes are financially necessary. Remember that identity development is a lifelong process - you don't need to have everything figured out immediately.

Perfectionism: The desire to get your new identity "right" can prevent experimentation and growth. Embrace the messiness of identity development and allow yourself to try things that might not work out.

Values-Based Decision Making

As you build a new identity, decision-making becomes both more complex and more important. Without the clear guidance of a stable professional identity, you need other frameworks for making choices about how to spend your time, energy, and attention.

Values Clarification: Regularly revisit and refine your understanding of what you value most deeply. These values can serve as decision-making criteria when career paths or opportunity choices feel unclear.

Purpose Alignment: Evaluate opportunities based on how well they align with your sense of purpose or mission rather than solely on financial or status considerations.

Growth Potential: Consider how different choices will support the identity development you're pursuing. Will this opportunity help you become more of who you want to be?

Integration Opportunities: Look for choices that allow you to integrate multiple aspects of your developing identity rather than forcing you to compartmentalize different parts of yourself.

Community Impact: Consider how your choices affect the communities you care about and want to be part of. Identity is partly about belonging and contribution to something larger than yourself.

Authenticity Assessment: Regularly check whether your choices and activities feel authentic to who you're becoming or whether you're making decisions based on outdated identity assumptions.

Building Resilient Identity Architecture

The goal of identity reconstruction isn't to find one perfect new identity that will never need to change again. Instead, the goal is to build what I call "resilient identity architecture" - a foundation of self-understanding that can adapt to changing circumstances while maintaining core stability.

Core vs. Adaptive Elements: Distinguish between core identity elements that remain stable (values, character strengths, fundamental purposes) and adaptive elements that can change based on circumstances (specific roles, skills, community affiliations).

Multiple Identity Anchors: Develop multiple sources of identity and meaning so that disruption to one area doesn't destabilize your entire sense of self.

Growth Orientation: Include adaptability, learning, and growth as core identity elements rather than seeing change as threatening to your identity.

Relationship Integration: Build identity around your capacity for relationship and community rather than just individual achievements or capabilities.

Service Orientation: Include contribution to others and to causes larger than yourself as central identity elements that provide meaning independent of economic circumstances.

Temporal Integration: Develop identity narratives that honor your past, engage fully with the present, and remain open to future possibilities.

The identity you build for an AI-integrated world won't just help you survive technological disruption - it can provide the foundation for a more fulfilling, authentic, and resilient approach to life. By moving beyond narrow professional identities toward more multifaceted, values-based self-concepts, you can not only adapt to an AI-dominated economy but potentially discover aspects of yourself and possibilities for contribution that you never explored when your identity was constrained by traditional career categories.

In our next chapter, we'll turn from the personal work of identity reconstruction to the practical challenge of financial survival in an AI-disrupted economy. While developing a robust identity provides psychological foundation for navigating change, you also need concrete strategies for maintaining financial stability while making the transition to new forms of work and value creation.

9

FINANCIAL SURVIVAL STRATEGIES

"It's not how much money you make, but how much money you keep, how hard it works for you, and how many generations you keep it for."
– Robert Kiyosaki

When Marcus Chen lost his financial analyst position to AI automation in March 2024, he had exactly $2,847 in his checking account, $18,000 in his 401(k), and monthly expenses of $3,200. Like many people who had built their careers around stable employment, he had never seriously considered what would happen if his expertise became obsolete overnight. His financial planning had been based on the assumption of steady income growth, predictable career advancement, and eventual retirement at 65.

"I kept thinking someone would recognize that the AI was missing important insights that only human analysts could provide," Marcus told me six months later. "I spent the first two months of unemployment sending out resumes for similar positions, convinced this was temporary. By month three, when my savings were nearly gone and no one was hiring human analysts, I realized I needed a completely different approach to money."

Marcus's experience illustrates one of the most challenging aspects of the AI revolution: traditional financial planning strategies, built around assumptions of stable employment and predictable career trajectories, may be inadequate for a world where AI can rapidly reshape entire industries. This chapter explores practical strategies for achieving financial stability and building wealth in an economy where human economic value is being redefined. The goal isn't just survival - it's developing financial strategies that provide security and opportunity while maintaining the flexibility to adapt as AI continues transforming the economic landscape.

Rethinking Financial Security

Traditional financial planning advice assumes a relatively stable economic environment where consistent employment, predictable salary increases, and conventional retirement savings can provide financial security. AI disruption challenges these assumptions and requires new approaches to financial planning that account for increased uncertainty and rapid change.

From Linear to Adaptive Planning: Instead of planning for a linear career progression with steady income increases, develop financial strategies that can adapt to multiple possible scenarios: career changes, income volatility, new types of work, and economic disruption.

Multiple Income Stream Development: Rather than relying on a single salary, build multiple income streams that provide both diversification and resilience. This might include part-time work, freelancing, passive income from investments, rental properties, or small business ventures.

Skill-Based Asset Building: Invest in developing skills and capabilities that can generate income across different economic scenarios rather than just climbing the ladder within a single industry or company.

Network as Financial Asset: Recognize that professional and personal networks represent valuable financial assets that can provide opportunities, referrals, support during transitions, and collaborative ventures.

Flexibility Premium: Value financial arrangements that provide flexibility and optionality, even if they offer lower immediate returns. The ability to adapt quickly to changing circumstances may be worth more than maximizing current income.

Dr. Elizabeth Warren's research on financial stability reveals that the families who weather economic disruption most successfully are those with diversified resources - multiple income sources, strong social networks, diverse skills, and financial cushions that provide time to adapt to changing circumstances.

Emergency Fund Reimagined

The traditional recommendation of saving three to six months of expenses in an emergency fund may be inadequate for AI-disrupted careers. When career transitions might take longer than expected and involve retraining or industry changes, emergency funds need to be larger and more strategically constructed.

Extended Timeline Planning: Instead of planning for short-term unemployment, prepare for potential career transitions that might take 12-18 months. This includes time for retraining, network building, and establishing new income streams.

Transition Fund vs. Emergency Fund: Create separate funds for different types of financial challenges. An emergency fund covers unexpected expenses; a transition fund provides resources for deliberate career changes, skill development, or business ventures.

Liquid Accessibility: Ensure that emergency funds are easily accessible without penalties or delays. High-yield savings accounts, money market funds, or short-term CDs provide both liquidity and modest returns.

Income Replacement Strategy: Instead of saving a fixed dollar amount, target saving enough to replace your income for an extended period while you develop new revenue streams. This might mean saving 12-18 months of expenses rather than the traditional 3-6 months.

Skill Development Fund: Allocate part of your emergency savings specifically for education, training, certifications, or other skill development that might be necessary during career transitions.

Maria Santos, the former customer service representative, learned the importance of extended financial preparation: "When my job was automated, I thought I'd find something similar quickly. Instead, I ended up spending eight months learning new skills, building a network in a different industry, and starting a small consulting business. My traditional three-month emergency fund would have been completely inadequate. I wish I'd planned for a longer, more complex transition."

Income Diversification Strategies

One of the most effective ways to build financial resilience in an AI-disrupted economy is developing multiple income streams that aren't all vulnerable to the same types of automation or economic disruption.

The Portfolio Career Approach: Instead of one full-time job, combine multiple part-time income sources. This might include consulting work, freelancing, part-time employment, teaching, and small business activities.

Skill Monetization: Identify ways to monetize skills you already possess through tutoring, coaching, consulting, or freelance work.

These activities can start as side income while you're employed and potentially grow into primary income sources.

Digital Product Creation: Develop digital products that can generate passive income over time. This might include online courses, e-books, templates, software tools, or subscription services related to your expertise.

Service-Based Business Development: Start small service-based businesses that leverage your human capabilities. These might include personal services, consulting, training, or specialty services that are difficult to automate.

Investment Income Building: Gradually build investment portfolios that can generate dividends, interest, or rental income to reduce dependence on employment income.

Gig Economy Participation: Strategically participate in gig economy activities that provide flexibility and income diversification while building skills and networks.

Jennifer Walsh, the financial advisor, successfully diversified her income during the transition to AI-enhanced financial services: "I kept my main advisory practice but added financial education workshops, a monthly investment newsletter, some freelance writing about personal finance, and a small real estate investment. When robo-advisors started competing with traditional advisory services, I had multiple revenue streams and wasn't dependent on any single one."

Skill-Based Investment Strategy

In an AI-dominated economy, investing in your own skills and capabilities may provide better returns than traditional financial investments. This requires thinking strategically about which skills to develop and how to monetize them effectively.

High-Value Human Skills: Invest in developing skills that are difficult to automate and likely to remain valuable. These include complex problem-solving, creativity, emotional intelligence, leadership, and specialized expertise in areas requiring human judgment.

Cross-Domain Capabilities: Develop skills that can be applied across multiple industries or contexts. Communication, project management, analytical thinking, and relationship building provide flexibility as different sectors evolve.

Technology Integration Skills: Learn to work effectively with AI tools and systems without trying to compete with them. This includes skills in directing AI systems, evaluating AI outputs, and integrating AI capabilities with human judgment.

Meta-Skills Development: Invest in learning how to learn, adapting to change, and building networks. These meta-skills help you acquire new capabilities as needed rather than becoming obsolete when specific skills are automated.

Credential and Certification Strategy: Selectively pursue credentials that provide access to protected professions or demonstrate competence in high-value areas. Focus on certifications that represent human judgment rather than just technical knowledge.

Teaching and Knowledge Transfer: Develop the ability to teach and transfer knowledge to others. As AI handles more routine tasks, the ability to help others adapt and learn becomes increasingly valuable.

Alternative Wealth Building Approaches

Traditional wealth building through employment income, employer-sponsored retirement plans, and conservative investing may be insufficient for financial security in an AI-disrupted economy. Alternative approaches can provide additional pathways to building wealth.

Entrepreneurial Wealth Building: Start small businesses that can grow over time and potentially be sold for significant returns. Focus on businesses that leverage human capabilities and provide value that's difficult to automate.

Real Estate Investment: Consider real estate investments that provide both rental income and potential appreciation. Look for properties that serve markets less likely to be disrupted by AI, such as housing for essential workers or properties in communities with strong local economies.

Intellectual Property Development: Create intellectual property - books, courses, methods, systems - that can generate ongoing royalties or licensing income.

Peer-to-Peer Investment: Explore peer-to-peer lending, crowdfunding investments, or other alternative investment platforms that can provide returns uncorrelated with traditional markets.

Community Investment: Invest in or create cooperative businesses, community-supported agriculture, local currencies, or other community-based economic activities that provide both financial returns and local economic resilience.

Skills-Based Partnership: Form partnerships or cooperatives with others who have complementary skills to create businesses or investment opportunities that none of you could pursue individually.

Debt Management in Uncertain Times

Managing debt becomes more critical when income is uncertain and career transitions are likely. High debt levels can prevent the flexibility needed to adapt to AI-driven economic changes.

Debt Prioritization: Focus first on paying down high-interest debt that doesn't provide tax benefits or investment potential. Credit card debt, personal loans, and other high-interest obligations should be eliminated as quickly as possible.

Strategic Debt Evaluation: Evaluate whether debt is supporting wealth building (such as mortgages on appreciating property or education loans that increase earning potential) or consuming resources without providing future benefits.

Emergency Credit Access: Maintain access to emergency credit through home equity lines of credit or other low-cost borrowing options, but avoid using these unless necessary.

Income-Based Repayment: For student loans and other debts with flexible repayment options, consider income-based repayment plans that can adjust during periods of lower income.

Debt Consolidation Strategy: Consolidate high-interest debt into lower-cost options where possible, but avoid extending repayment periods unnecessarily.

Avoid Lifestyle Inflation: Resist the temptation to increase debt levels during periods of higher income. Instead, use income increases to build financial resilience and reduce debt burdens.

Building Multiple Revenue Streams

Creating multiple revenue streams provides both financial stability and opportunities for growth. The key is developing streams that complement each other while reducing overall risk.

Primary Income Stabilization: Maintain at least one relatively stable income source while developing others. This might be part-time employment, consulting contracts, or established freelance relationships.

Passive Income Development: Gradually build income sources that require minimal ongoing time investment. These might include rental properties, dividend-paying investments, royalties from creative works, or affiliate marketing income.

Scalable Service Development: Create service offerings that can grow over time without proportional increases in your time investment. This might include online courses, group coaching programs, or subscription services.

Seasonal Income Optimization: Develop income sources that peak at different times of year to smooth out seasonal variations. Tax preparation services, holiday retail, summer coaching programs, or winter consulting might complement each other.

Geographic Income Diversification: Explore income opportunities that serve different geographic markets to reduce dependence on local economic conditions.

Currency and Economic Diversification: Consider income sources denominated in different currencies or tied to different economic sectors to reduce correlation risk.

Robert Chen, the accountant who transitioned from traditional tax preparation to financial education and small business consulting, exemplifies successful revenue stream development: "I realized I needed to stop thinking about one job and start thinking about multiple ways to generate income. I do tax work during tax season, teach financial literacy classes year-round, consult with small businesses on financial systems, and created an online course about bookkeeping for freelancers. No single stream provides all my income, but together they give me more total income and much more security than my old accounting job ever did."

Investment Strategies for Uncertain Times

Traditional investment advice - diversified stock and bond portfolios held for the long term - may need modification for an AI-disrupted economy where some industries could disappear while others emerge rapidly.

Sector Diversification with AI Awareness: Diversify investments across sectors while being mindful of which industries are most vulnerable to AI disruption. Avoid over-concentrating investments in sectors likely to be heavily automated.

Human-Centric Investment Focus: Consider investments in companies and sectors that benefit from AI adoption while providing essential human services. Healthcare, education, elder care, and community services may benefit from AI efficiency while requiring human delivery.

Infrastructure and AI-Enabling Investments: Invest in the infrastructure that supports AI development and deployment - data centers, semiconductors, renewable energy, and telecommunications.

Small and Local Business Investment: Support local businesses and community development through investment in companies that serve local markets and provide services requiring human interaction.

Education and Skill Development Investment: Invest in companies and organizations that help people develop skills for an AI economy - educational technology, professional development, and retraining services.

Flexibility-Preserving Investment Strategy: Maintain enough liquidity to take advantage of opportunities or manage unexpected challenges. This might mean holding more cash or short-term investments than traditional advice suggests.

Financial Planning for Career Transitions

AI disruption often requires career transitions that can take months or years to complete successfully. Financial planning must account for these extended transition periods and the investments they require.

Transition Timeline Planning: Develop realistic timelines for career transitions that account for skill development, network building, credential acquisition, and market entry. Plan financially for transitions that might take 12-24 months.

Skill Development Budget: Allocate resources specifically for education, training, certifications, and other skill development activities. This might include tuition, books, conference attendance, and mentorship programs.

Network Building Investment: Budget for activities that build professional networks - conference attendance, professional association memberships, informational interviews, and relationship-building activities.

Trial Period Planning: Allow for extended trial periods when exploring new career directions. This might include unpaid internships, volunteer work, or low-paid apprenticeships that provide experience and network access.

Bridge Income Strategy: Develop sources of bridge income that can support you during career transitions while you build new professional capabilities and relationships.

Opportunity Fund Creation: Maintain resources that can be deployed quickly to take advantage of unexpected opportunities - business purchases, partnership investments, or educational programs with limited enrollment periods.

Technology and Financial Management

AI tools can actually help with financial management and planning, even as AI disrupts traditional careers. Learning to use these tools effectively can improve financial outcomes while freeing time for higher-value activities.

AI-Assisted Budget Management: Use AI-powered budgeting apps that can track expenses, identify patterns, and suggest optimizations based on your financial goals and spending behavior.

Investment Management Tools: Leverage robo-advisors and AI-powered investment platforms for portfolio management while maintaining human oversight of major financial decisions.

Tax Optimization: Use AI tax software that can identify deductions, credits, and strategies you might miss while maintaining human review of complex situations.

Financial Planning Analysis: Employ AI tools that can model different financial scenarios and help you understand the implications of various financial decisions.

Income Optimization: Use AI platforms that can help identify freelance opportunities, optimize pricing for services, or match your skills with market demands.

Risk Assessment: Leverage AI tools that can assess and monitor various types of financial risk while maintaining human judgment about acceptable risk levels.

Building Long-Term Wealth

While managing immediate financial stability is crucial, building long-term wealth requires strategies that can compound over time and provide financial independence regardless of employment status.

Asset Acquisition Strategy: Focus on acquiring assets that can appreciate over time and generate income - real estate, businesses, intellectual property, and investment portfolios.

Compound Growth Maximization: Start building long-term investments as early as possible to maximize compound growth over time. Even small amounts invested consistently can grow significantly over decades.

Tax-Advantaged Account Optimization: Maximize contributions to 401(k)s, IRAs, HSAs, and other tax-advantaged accounts that can reduce current taxes while building future wealth.

Business Equity Building: Consider starting or investing in businesses that can grow in value over time and potentially be sold for significant returns.

Inflation Protection: Include investments that can maintain purchasing power during inflationary periods - real estate, commodities, or inflation-protected securities.

Legacy and Estate Planning: Develop estate planning strategies that can protect and transfer wealth to future generations while minimizing tax implications.

Community and Cooperative Financial Strategies

Individual financial planning may be insufficient for navigating AI-driven economic disruption. Community-based and cooperative financial strategies can provide additional resources and risk-sharing opportunities.

Community Investment Cooperatives: Participate in or create investment cooperatives that pool resources to make larger investments or support community businesses.

Skill-Sharing Networks: Join or create networks where people exchange services and skills rather than always using money. These networks can reduce expenses while building relationships.

Group Purchasing Power: Combine purchasing power with others to reduce costs for insurance, bulk purchases, or shared services.

Mutual Aid Networks: Participate in communities that provide mutual support during financial difficulties, career transitions, or other challenges.

Community Development Finance: Invest in community development financial institutions, credit unions, or other organizations that support local economic development.

Cooperative Business Development: Consider forming cooperative businesses with others who have complementary skills and shared values.

Financial survival in an AI-disrupted economy requires more than just traditional financial planning. It demands new strategies that account for increased uncertainty, longer career transitions, and the need to build multiple income streams while developing human capabilities that can't be automated. The key is to start early, diversify broadly, and maintain flexibility while building the financial foundation that can support both stability and opportunity in an economy where artificial intelligence is reshaping how value is created and distributed.

In our next chapter, we'll shift focus from individual financial strategies to the broader systemic changes needed in government and corporate policies to support widespread adaptation to AI disruption. While personal financial planning is crucial, systemic challenges require collective solutions that go beyond what individuals can achieve alone.

PART III: THE REBUILD

10

GOVERNMENT, CORPORATIONS, AND RESPONSIBILITY

"The ultimate measure of a man is not where he stands in moments of comfort and convenience, but where he stands at times of challenge and controversy." – Martin Luther King Jr.

In March 2024, the city of Akron, Ohio, became an unlikely epicenter of the debate over AI responsibility when Consolidated Insurance announced it would be automating 1,200 customer service and claims processing jobs over the next eighteen months. The company's press release emphasized the benefits: faster claim processing, 24/7 availability, and reduced costs that could be passed on to customers. What the press release didn't mention was that Consolidated had received $3.2 million in tax incentives from the city just two years earlier to expand their operations and "create jobs for the community."

Mayor Patricia Williams called an emergency town hall meeting that drew over 800 residents - mostly employees facing job loss and their families. The anger was palpable, but it wasn't just directed at Consolidated Insurance. People were equally frustrated with a system that seemed to privatize the benefits of AI automation while socializing the costs. "They got tax breaks from our community to create jobs," said

James Morrison, a claims processor who had worked at Consolidated for twelve years. "Now they're using AI to eliminate those same jobs and boost their profits. Meanwhile, my family loses health insurance, I can't make my mortgage payments, and the city loses tax revenue when I have to move away to find work. Where's the responsibility for the community that supported them?"

The Akron situation illustrates one of the most critical challenges of the AI revolution: while individual adaptation strategies are essential, the scale and speed of AI disruption requires coordinated responses from governments and corporations. The question isn't whether AI adoption should continue - that train has left the station. The question is whether we can develop systems of shared responsibility that help communities and individuals navigate the transition while ensuring that the benefits of AI advancement are distributed more equitably. This chapter examines what responsible AI adoption looks like from institutional perspectives and what individuals can do to advocate for better policies and corporate practices that support widespread adaptation rather than leaving everyone to fend for themselves.

The Current Responsibility Gap

The rapid deployment of AI systems has created what economists call an "externality problem" - the costs and benefits of AI adoption are not borne by the same parties. Companies that implement AI systems capture most of the efficiency gains and cost reductions, while workers, communities, and government systems bear most of the transition costs.

Corporate Benefit Concentration: Companies implementing AI typically see immediate benefits: reduced labor costs, increased processing speed, improved consistency, and enhanced competitive positioning. These benefits flow directly to shareholders and executives while the costs of displacement fall elsewhere.

Distributed Social Costs: The costs of AI displacement are distributed across multiple parties: displaced workers lose income and face retraining costs, families experience stress and potential relocation, communities lose tax revenue and economic activity, and government systems face increased demand for unemployment benefits, retraining programs, and social services.

Timeline Mismatch: Companies can implement AI systems rapidly to capture competitive advantages, but individuals and communities need much longer timeframes to adapt successfully. This creates a temporal mismatch where costs are immediate but adaptation takes months or years.

Information Asymmetry: Companies have detailed information about their AI implementation timelines and impact on employment, but workers and communities often receive little advance notice or information to help them prepare for changes.

Market Failure: Traditional market mechanisms don't account for the social costs of rapid technological displacement, creating situations where economically rational decisions for individual companies create significant costs for society as a whole.

Dr. Erik Brynjolfsson's research at MIT reveals that this responsibility gap isn't accidental - it's a predictable result of how AI development and deployment incentives are currently structured. "We have economic systems that reward rapid AI adoption without accounting for transition costs," he explains.

> **"Until we align incentives better, we'll continue seeing situations where AI creates tremendous value while leaving many people worse off."**

Corporate Responsibility Models

Some forward-thinking companies have begun developing models for responsible AI adoption that account for broader stakeholder impacts rather than focusing solely on shareholder returns. These models provide templates for how businesses can implement AI while supporting rather than abandoning affected communities.

The Partnership Approach

Companies using this model treat AI implementation as an opportunity to enhance human capabilities rather than simply replace workers. Instead of viewing employees as costs to be minimized, they see them as assets to be augmented and redeployed.

Retraining and Upskilling: Companies invest significantly in helping current employees develop skills that complement AI systems rather than compete with them. This includes both technical training and development of uniquely human capabilities like emotional intelligence, creative problem-solving, and relationship management.

Role Evolution Rather Than Elimination: Instead of eliminating positions, companies redesign roles to focus on tasks that benefit from human-AI collaboration. Customer service representatives might become customer relationship specialists who handle complex issues while AI manages routine inquiries.

Internal Mobility Programs: Companies create pathways for employees in AI-affected positions to move into other roles within the organization, often with enhanced responsibilities that leverage both their existing experience and new AI-augmented capabilities.

Extended Transition Timelines: Rather than implementing AI as quickly as possible, companies adopt longer transition timelines that allow for adequate training, role development, and adjustment periods.

Microsoft's approach to implementing AI in their customer support operations exemplifies this model. Instead of replacing support staff with chatbots, they trained existing employees to work alongside AI systems, handling escalated cases that require human judgment while AI manages routine troubleshooting. Employee satisfaction actually increased as workers focused on more challenging and meaningful interactions.

The Community Investment Model

Some companies recognize that their long-term success depends on healthy communities and invest in community development as part of their AI implementation strategy.

Local Retraining Partnerships: Companies partner with local educational institutions, community colleges, and training organizations to provide retraining opportunities not just for their own employees but for community members facing AI displacement from other employers.

Economic Development Investment: Companies invest in local economic development initiatives that can create new employment opportunities as traditional jobs are automated. This might include supporting startup incubators, funding community infrastructure projects, or investing in local suppliers.

Gradual Implementation with Community Support: Companies implement AI systems gradually while providing extensive support for affected workers and communities, including extended severance packages, job placement assistance, and community development investments.

Transparency and Communication: Companies provide detailed information about AI implementation timelines and impacts, allowing workers and communities to prepare for changes rather than being surprised by sudden announcements.

Success Metric Expansion: Companies measure success not just by cost reduction and efficiency gains but also by community impact metrics, employee satisfaction, and long-term sustainability indicators.

The Shared Value Creation Model

This approach recognizes that AI's greatest value comes from combining human and artificial intelligence rather than substituting one for the other.

Human-AI Collaboration Design: Systems are designed from the beginning to optimize human-AI collaboration rather than human replacement. This creates new types of jobs that didn't exist before while eliminating others.

Innovation Investment: Companies invest heavily in research and development that creates new products, services, and business models that require both AI capabilities and human skills.

Customer Experience Enhancement: AI is used to handle routine aspects of customer interaction while human employees focus on relationship building, problem-solving, and providing the empathy and understanding that customers value.

Employee Ownership Programs: Companies provide employees with ownership stakes in AI-driven productivity gains, ensuring that workers benefit from rather than just bear the costs of technological advancement.

Stakeholder Governance: Companies include employee representatives, community members, and other stakeholders in governance processes related to AI implementation and strategic planning.

Government Policy Frameworks

While corporate responsibility initiatives are encouraging, they remain voluntary and inconsistent. Government policy frameworks can create systematic approaches to managing AI disruption while ensuring that the benefits and costs are distributed more fairly across society.

Universal Basic Income and Variations

The most discussed government response to AI displacement is Universal Basic Income (UBI) - providing all citizens with regular, unconditional cash payments sufficient to meet basic needs regardless of employment status.

Pilot Program Results: UBI pilot programs in Finland, Kenya, Stockton California, and other locations have shown mixed but generally positive results. Recipients report reduced financial stress, improved physical and mental health, and increased ability to pursue education, caregiving, or entrepreneurial activities.

Implementation Challenges: Large-scale UBI implementation faces significant challenges including funding requirements, potential inflation effects, political feasibility, and questions about work incentives and social cohesion.

Targeted Variations: Some economists propose targeted variations like Universal Basic Assets (providing everyone with capital to invest), job guarantees that provide employment in socially beneficial activities, or negative income tax systems that provide support inversely related to income levels.

Funding Mechanisms: Proposed funding mechanisms include taxes on AI and automation, carbon taxes, land value capture, financial transaction taxes, or reallocation of existing social spending.

Alaska's Permanent Fund Dividend, funded by oil revenues and providing annual payments to all residents, offers one model for how resource-based UBI systems might function, though the amounts are currently too small to provide basic income security.

Retraining and Education Investment

Government investment in massive retraining and education programs can help workers adapt to AI-changed labor markets while building the human capital needed for economic growth.

Community College Partnerships: Federal and state investments in community college programs that provide rapid retraining for displaced workers, focusing on skills that complement rather than compete with AI systems.

Portable Benefits Systems: Creating benefit systems that aren't tied to specific employers, allowing workers to maintain healthcare, retirement savings, and other benefits while transitioning between different types of work.

Lifelong Learning Accounts: Government-supported individual accounts that workers can use throughout their careers for education, training, and skill development as technological change requires adaptation.

Industry-Specific Transition Programs: Targeted programs that help workers in specific industries - manufacturing, retail, financial services - transition to new roles as those industries are transformed by AI.

Rural and Community Focus: Special attention to rural and smaller communities that may have fewer resources for managing AI transitions and less diverse economic bases to absorb displaced workers.

Regulation and Transition Management

Government regulation can manage the pace and process of AI implementation to allow for more orderly transitions while still enabling innovation and economic growth.

AI Impact Assessment Requirements: Requiring companies to assess and report on the employment and community impacts of AI implementations, similar to environmental impact assessments for major projects.

Transition Timeline Regulation: Establishing minimum timelines for major AI implementations that affect employment, ensuring that workers and communities have adequate time to prepare for changes.

Worker Notification Requirements: Expanding existing plant closure notification laws to cover AI-driven job displacement, giving workers and communities advance warning of technological changes.

Severance and Support Standards: Establishing minimum standards for severance packages, retraining support, and transition assistance when companies implement job-displacing AI systems.

Community Impact Mitigation: Requiring companies to contribute to community development funds or provide other forms of mitigation when AI implementations significantly impact local employment.

Tax Policy and Incentive Alignment

Tax policy can be used to better align corporate incentives with community needs while funding transition support and social investment.

Automation Taxes: Taxes on companies that replace human workers with AI systems, with revenues used to fund retraining programs, community development, and social support for displaced workers.

Tax Incentive Reform: Modifying existing tax incentives to reward companies for responsible AI implementation that includes worker retraining and community support rather than just job creation or economic development.

Corporate Social Investment Credits: Tax credits for companies that invest significantly in worker retraining, community development, and responsible AI implementation practices.

Graduated Implementation Incentives: Tax incentives that reward gradual, well-managed AI implementation over rapid deployment that maximizes displacement.

Community Benefit Requirements: Requiring companies that receive public benefits - tax incentives, infrastructure investment, regulatory advantages - to provide community benefits when they implement job-displacing technologies.

International and Federal Policy Coordination

AI disruption is a global phenomenon that requires coordination across different levels of government and between nations to manage effectively.

Federal Standards and Support: Establishing national standards for responsible AI implementation while providing federal support for communities and workers affected by technological displacement.

Interstate Cooperation: Developing mechanisms for states to coordinate responses to AI disruption, sharing best practices and resources for worker retraining and economic transition.

International Trade and Labor Standards: Including provisions for responsible AI implementation and worker protection in international trade agreements and labor standards.

Cross-Border Worker Mobility: Creating programs that allow workers displaced by AI in one region to access opportunities and support in other areas with labor shortages or economic growth.

Global Technology Governance: Participating in international efforts to develop governance frameworks for AI development and deployment that account for social and economic impacts.

The Role of Labor Organizations

Labor unions and worker organizations play crucial roles in advocating for responsible AI implementation and supporting workers through technological transitions.

Collective Bargaining Evolution: Unions are developing new approaches to collective bargaining that address AI implementation, including provisions for advance notification, retraining support, and shared benefits from productivity gains.

Cross-Industry Organization: As AI affects workers across many industries, labor organizations are developing new forms of cross-industry cooperation and worker solidarity that transcend traditional union boundaries.

Political Advocacy: Labor organizations advocate for government policies that support workers through AI transitions while pushing for corporate accountability in technological implementation.

Worker Education and Support: Unions provide education about AI impacts, rights, and adaptation strategies while offering emotional and practical support for members facing technological displacement.

Alternative Economic Models: Some labor organizations are exploring alternative economic models like worker cooperatives, platform cooperatives, and other structures that give workers more control over technological implementation in their workplaces.

Corporate Accountability Mechanisms

Creating accountability for corporate AI implementation requires new mechanisms that go beyond traditional market-based approaches.

Stakeholder Governance Requirements: Policies that require companies above certain sizes to include employee representatives, community members, or other stakeholders in governance processes related to major technological implementations.

Social Impact Reporting: Requirements for companies to report annually on the social and economic impacts of their AI implementations, including effects on employment, communities, and inequality.

Community Benefit Agreements: Voluntary or required agreements between companies and communities that specify benefits and protections when companies implement major technological changes.

Worker Protection Standards: Establishment of industry standards for worker protection during AI implementation, including retraining, transition support, and benefit continuation.

Transparency Requirements: Policies requiring companies to provide clear information about AI implementation timelines, expected impacts, and mitigation measures to affected workers and communities.

Success Stories and Model Programs

Several communities and organizations have developed innovative approaches to managing AI transitions that provide models for broader implementation.

The Danish Model

Denmark's approach to managing technological transitions provides one model for coordinated government, business, and labor cooperation.

Flexicurity System: Denmark combines flexible labor markets with generous unemployment benefits and extensive retraining programs, allowing workers to transition between jobs without experiencing severe economic hardship.

Tripartite Cooperation: Government, employers, and labor unions work together to anticipate technological changes and develop coordinated responses that protect workers while enabling innovation.

Lifelong Learning Investment: Massive public investment in adult education and retraining programs that are available to all workers throughout their careers.

Social Safety Net Integration: Comprehensive social safety nets that provide healthcare, housing support, and income security independent of employment status.

Regional Innovation Initiatives

Some regions have developed innovative approaches to managing AI transitions while building new economic opportunities.

Pittsburgh's Robotics Transition: Pittsburgh's transformation from a steel manufacturing center to a robotics and AI research hub demonstrates how regions can build new economies around technological change rather than just adapting to it.

Singapore's SkillsFuture Program: Singapore provides all citizens with credits for lifelong learning and skills development, creating a national system for continuous adaptation to technological change.

Germany's Industry 4.0 Initiative: Germany's coordinated approach to integrating AI and automation in manufacturing includes extensive worker training, social partnership, and gradual implementation timelines.

Advocacy and Individual Action

While systemic change requires institutional action, individuals can play important roles in advocating for responsible AI implementation and better policies to support community adaptation.

Community Organization: Participating in or organizing community groups that advocate for responsible AI implementation by local employers and better government policies to support workers and families.

Political Engagement: Voting for candidates who support policies that address AI disruption and contacting elected representatives about the need for proactive policies to manage technological transitions.

Corporate Engagement: As consumers, employees, and community members, individuals can pressure companies to adopt more responsible approaches to AI implementation through consumer choices, shareholder advocacy, and public pressure.

Information Sharing: Helping to educate others about AI impacts and policy options while building community awareness and engagement around these issues.

Coalition Building: Working with labor organizations, community groups, faith communities, and other organizations to build coalitions that can advocate effectively for better policies and practices.

Model Development: Participating in pilot programs, cooperative ventures, and other experiments that can demonstrate alternative approaches to managing AI transitions.

The Path Forward

Creating systems of shared responsibility for AI transitions requires coordinated action across multiple levels and institutions. The goal isn't to prevent AI adoption but to ensure that its benefits are shared more equitably while its costs don't fall disproportionately on those least able to bear them.

Immediate Actions: Expanding unemployment benefits, increasing funding for retraining programs, and requiring better corporate disclosure of AI implementation plans.

Medium-term Reforms: Developing new social insurance systems, creating portable benefits, and establishing standards for responsible AI implementation.

Long-term Transformation: Building economic systems that better distribute the benefits of technological advancement while maintaining incentives for innovation and growth.

Community Resilience: Strengthening local communities and regional economies so they're better able to adapt to technological change while maintaining social cohesion and economic opportunity.

The challenge of responsible AI implementation is ultimately about the kind of society we want to create with these powerful new technologies. Will AI benefits flow primarily to those who own the technology while costs fall on everyone else? Or can we develop systems that harness AI's capabilities while ensuring that technological advancement serves broader human flourishing? The answers to these questions won't emerge automatically from market forces or technological development alone. They require deliberate choices about policies, corporate practices, and social institutions that prioritize shared prosperity and community resilience alongside innovation and economic growth.

In our next chapter, we'll explore how communities can strengthen their social fabric and build stronger connections in an increasingly digital world, recognizing that technology is only as valuable as the human relationships and social bonds it serves.

11

COMMUNITY AND CONNECTION IN THE DIGITAL AGE

"If you want to go fast, go alone. If you want to go far, go together." – African Proverb

When the Riverside Community Center in Portland, Oregon, lost its funding in 2023, longtime neighborhood resident Carmen Rodriguez feared that one of the few remaining gathering places in her changing community would disappear forever. The center had served as more than just a building - it was where neighbors met during the weekly farmers market, where parents connected while their children played, where job seekers shared leads and retirees offered wisdom, where the community came together during celebrations and crises.

But something unexpected happened when the center's closure was announced. Instead of accepting the loss, neighbors began organizing in ways that surprised even longtime community activists. Sarah Chen, a software developer who had been working remotely since AI automated much of her previous role, offered to coordinate online communications. Marcus Williams, a former retail manager whose job was eliminated by automated checkout systems, had time to canvas the neighborhood and organize meetings. Elena Santos, whose administrative position was restructured around AI tools, used her newly flexible schedule to research funding options and write grant applications. Within six months, this unlikely group of neighbors - many of whom had initially felt displaced and discouraged by technological change - had not only saved the community center but transformed it into a hub for mutual support, skill sharing, and collective adaptation to their rapidly changing economic landscape.

> **"Losing our jobs to AI was devastating at first," Carmen told me a year later. "But it also gave us something we hadn't had in years: time to actually connect with each other and work on things that mattered to our community. We discovered that while AI could replace our work tasks, it couldn't replace the satisfaction of solving problems together and taking care of each other."**

The Portland story illustrates one of the most hopeful aspects of navigating AI disruption: while technology can fragment communities and isolate individuals, it can also create opportunities for deeper connection and more meaningful forms of community engagement. The key is being intentional about building and maintaining the social bonds that technology cannot provide.

The Community Deficit in Digital Society

Before exploring solutions, it's important to understand how digital technologies - including but not limited to AI - have contributed to what sociologist Robert Putnam calls "bowling alone" syndrome: the decline of social capital and community connection that has been documented across much of the developed world.

Physical Gathering Space Decline: Many traditional community gathering places - local shops, community centers, libraries, religious institutions - have seen reduced use as people increasingly meet their needs through online platforms and digital services.

Social Interaction Substitution: Digital communication often substitutes for face-to-face interaction in ways that provide information exchange but less emotional connection, empathy development, and community bonding.

Geographic Dispersion: Remote work and digital services enable people to live farther from employment centers and service providers, potentially reducing daily interactions with neighbors and local community members.

Attention Fragmentation: Constant digital connectivity can fragment attention and reduce the sustained focus that deep relationships and community engagement require.

Economic Community Disruption: As local businesses are displaced by online retailers and AI-powered services, communities lose not just economic activity but also the informal social networks that form around local commerce.

Skill and Knowledge Centralization: When AI systems centralize capabilities that were once distributed across community members - from directions and recommendations to problem-solving and decision-making - communities may lose some of the interdependence that creates social bonds.

However, these challenges are not inevitable consequences of technology but rather results of how we've chosen to implement and engage with digital tools. The same technologies that can isolate can also connect, and the same AI systems that can replace human functions can also free people to engage more meaningfully with each other.

Digital Tools for Community Building

When used intentionally, digital technologies can actually strengthen rather than weaken community connections by making it easier to coordinate, communicate, and collaborate around shared interests and challenges.

Neighborhood Communication Platforms: Apps like Nextdoor, local Facebook groups, and community-specific platforms can help neighbors stay informed about local issues, coordinate mutual aid, and organize community activities.

Event Coordination and Participation: Digital tools make it easier to organize community events, coordinate volunteer activities, and ensure that community members know about opportunities for engagement.

Skill and Resource Sharing: Online platforms can help community members identify and share skills, tools, and resources with each other, creating economic connections that strengthen social bonds.

Crisis Response Coordination: During emergencies or community challenges, digital communication can enable rapid coordination of mutual aid, information sharing, and collective response.

Interest-Based Community Formation: Digital tools can help people with shared interests - from gardening to genealogy - find each other and organize activities, even in geographically dispersed communities.

Civic Engagement Enhancement: Digital platforms can increase participation in local government, community planning, and civic decision-making by making it easier for people to stay informed and contribute input.

The Portland community center organizers used a combination of these approaches. Sarah Chen created a neighborhood WhatsApp group for daily communication, a shared Google Drive for organizing documents and resources, and a simple website that helped neighbors learn about each other's skills and availability. These digital tools didn't replace face-to-face interaction - they facilitated and enhanced it.

Building Resilient Local Networks

AI disruption makes resilient local networks more important, not less. When traditional sources of economic security and social structure are changing rapidly, strong community networks provide crucial support for adaptation and mutual aid.

Economic Resilience Through Local Networks

Local Exchange Systems: Communities can develop local currency systems, time banks, or barter networks that keep economic activity local and build relationships through economic exchange.

Cooperative Business Development: Community members can pool resources to start cooperative businesses that serve local needs while providing employment and ownership opportunities for community members.

Skill Development Networks: Communities can organize skill-sharing networks where members teach each other capabilities ranging from practical skills to professional development, reducing dependence on formal education and training institutions.

Local Investment Circles: Community members can form investment clubs or crowdfunding networks to support local businesses and economic development projects.

Mutual Aid Networks: Organized systems for neighbors to help each other with childcare, elder care, home maintenance, transportation, and other needs can reduce costs while building relationships.

Social Resilience Through Community Connection

Inter-generational Programming: Creating opportunities for different age groups to interact and learn from each other helps communities preserve wisdom while adapting to change.

Cultural Preservation and Innovation: Communities can organize activities that preserve cultural traditions while also creating new forms of artistic and cultural expression.

Conflict Resolution Capacity: Building community capacity for resolving conflicts and making collective decisions helps neighborhoods handle challenges and changes more effectively.

Social Support Systems: Organized systems for emotional support, mental health assistance, and crisis intervention help community members navigate personal challenges while strengthening social bonds.

Civic Engagement Infrastructure: Creating regular opportunities for community members to participate in local decision-making and civic life builds democratic capacity and community ownership.

The New Volunteerism

AI disruption may actually create opportunities for expanded volunteerism and community service as more people have time available due to job displacement or increased efficiency in paid work.

Skills-Based Volunteering: People with professional skills can volunteer those capabilities to benefit community organizations, helping them become more effective while providing meaningful engagement for volunteers.

Community Problem-Solving: Volunteers can focus on identifying and addressing local problems that government and market solutions don't adequately address.

Peer Support and Mentoring: People who have successfully navigated career transitions, skill development, or other challenges can mentor others facing similar situations.

Community Education and Learning: Volunteers can organize and teach informal education programs that help community members develop new skills or explore interests.

Civic Innovation: Volunteers can work on improving local government services, community infrastructure, and civic processes.

Environmental Stewardship: Community volunteers can organize local environmental improvement projects that benefit everyone while building social connections.

Marcus Williams, the former retail manager from Portland, exemplifies this new volunteerism. After losing his job to automated systems, he initially felt worthless and directionless. But volunteering to help other displaced workers navigate job searches and career transitions gave him a sense of purpose and helped him discover skills he didn't know he possessed.

> **"I realized I was actually good at helping people think through their problems and figure out next steps," Marcus said. "The AI could process applications and match skills to job requirements, but it couldn't sit with someone who was scared and confused and help them believe in themselves again. That became my new calling."**

Intergenerational Wisdom Exchange

One of the most valuable community-building opportunities in an AI-disrupted world involves connecting different generations so they can learn from each other's experiences and perspectives.

Technology and Tradition Integration: Older community members can share life experience, practical skills, and historical perspective while younger members help with technology adoption and digital literacy.

Career Transition Mentoring: People who have successfully navigated multiple career changes throughout their lives can mentor others facing AI-related career disruption.

Life Skills Sharing: Older generations can teach practical skills like cooking, gardening, home maintenance, and financial management that remain valuable regardless of technological change.

Historical Context Provision: Understanding how communities have adapted to previous technological and economic disruptions can provide perspective and hope for current challenges.

Cultural Knowledge Transmission: Older community members can share cultural knowledge, stories, and traditions that provide identity and meaning beyond economic roles.

Purpose and Meaning Exploration: People who have lived through different life stages can help younger community members think about purpose, meaning, and life direction during times of uncertainty.

Community-Supported Adaptation

Just as Community Supported Agriculture (CSA) connects urban consumers directly with farmers, communities can develop systems for supporting members through major life and career transitions.

Transition Support Circles: Small groups of community members can commit to supporting each other through career changes, skill development, or other major life transitions.

Community Learning Cooperatives: Groups of neighbors can organize cooperative learning experiences, sharing the costs and coordination of skill development while building relationships.

Resource Sharing Networks: Communities can organize systems for sharing tools, equipment, vehicles, and other resources that reduce individual costs while creating opportunities for interaction.

Childcare and Elder Care Cooperatives: Community members can organize cooperative approaches to childcare and elder care that reduce costs while providing social interaction and mutual support.

Community Investment in Individual Development: Just as CSA members invest in farms, community members can invest in each other's skill development, business ventures, or creative projects.

Collective Purchasing Power: Communities can organize to purchase goods and services collectively, reducing costs while supporting local businesses.

The Role of Physical Spaces

While digital tools can enhance community building, physical spaces where people can gather, work, and interact remain crucial for developing the deep relationships that sustain communities through difficult transitions.

Community Centers and Libraries: Supporting and revitalizing community centers, libraries, and other public spaces where people can gather for various purposes.

Shared Working Spaces: Creating spaces where people can work independently while being around others, providing social interaction for remote workers and freelancers.

Community Gardens and Food Systems: Organizing community gardens, food cooperatives, and other food-related activities that bring people together around basic needs while building local resilience.

Maker Spaces and Repair Cafes: Creating spaces where people can work on projects, repair items, and share tools and skills.

Community Kitchens and Gathering Spaces: Spaces where people can cook and eat together, host community events, and gather for various purposes.

Outdoor Community Spaces: Parks, playgrounds, walking trails, and other outdoor spaces where people can interact informally while engaging in physical activity or recreation.

The Portland community center became successful partly because it served multiple functions: coworking space during the day, community meeting space in the evenings, workshop and skill-sharing space on weekends, and social gathering space for community events.

Digital-Physical Integration

The most effective community building approaches integrate digital and physical interaction rather than treating them as separate or competing modes of connection.

Digital Coordination of Physical Activities: Using digital tools to organize, coordinate, and promote face-to-face community activities and gatherings.

Online-to-Offline Relationship Building: Starting relationships or identifying common interests online, then transitioning to face-to-face interaction and collaboration.

Hybrid Event Formats: Community events that include both in-person and remote participation options, making it easier for people with different schedules or mobility constraints to participate.

Digital Documentation of Community Life: Using digital tools to document and share community activities, achievements, and stories in ways that build community identity and pride.

Virtual Community Spaces: Creating online spaces where community members can interact, share resources, and maintain connections between face-to-face gatherings.

Augmented Local Knowledge: Using digital tools to enhance rather than replace local knowledge sharing, such as apps that help people learn about local history, resources, or points of interest.

Community Leadership in the AI Age

Effective community building requires leadership, but the kind of leadership needed in an AI-disrupted world may be different from traditional models.

Facilitative Leadership: Leaders who focus on facilitating community member participation and decision-making rather than directing activities from the top down.

Distributed Leadership: Sharing leadership responsibilities across multiple community members rather than depending on single leaders or small groups.

Adaptive Leadership: Leaders who can help communities navigate uncertainty and change while maintaining focus on core values and relationships.

Bridge-Building Leadership: Leaders who can connect different groups within communities and help people find common ground across differences.

Inclusive Leadership: Ensuring that community leadership includes people from different backgrounds, age groups, and life experiences.

Technology-Fluent Leadership: Leaders who understand how to use digital tools effectively while maintaining focus on human relationships and community values.

Elena Santos, who helped save the Portland community center, exemplifies this new community leadership. Rather than taking charge and directing activities, she focused on helping neighbors identify their own interests and capabilities, then creating opportunities for people to contribute according to their strengths and availability.

Measuring Community Health

As communities work to strengthen social bonds and build resilience, it's important to track progress and identify areas that need attention.

Social Connection Indicators: Measuring the strength and frequency of social connections among community members through surveys, participation rates, and network analysis.

Civic Engagement Levels: Tracking participation in local government, community organizations, and volunteer activities.

Economic Resilience Metrics: Measuring local economic activity, business formation, employment stability, and economic diversity.

Cultural Vitality Indicators: Tracking cultural activities, artistic expression, celebration of diversity, and preservation of community traditions.

Community Problem-Solving Capacity: Assessing how effectively the community identifies and addresses local challenges and opportunities.

Inter-group Cooperation: Measuring cooperation and understanding between different demographic groups within the community.

Barriers to Community Building

Building stronger communities in the AI age faces several significant barriers that require deliberate attention and creative solutions.

Time Poverty: Many people feel too busy or stressed to engage in community activities, especially during periods of economic uncertainty or career transition.

Social Skill Atrophy: As digital communication becomes more common, some people may lose confidence in face-to-face social interaction and community engagement.

Economic Inequality: Growing economic inequality can create divisions within communities and make it harder for people to find common ground.

Geographic Mobility: High levels of residential mobility can make it harder to build lasting community relationships and institutional memory.

Cultural Fragmentation: Communities may struggle with increasing cultural diversity or generational differences if they lack frameworks for building bridges across differences.

Institutional Decline: The decline of traditional community institutions - religious organizations, unions, civic groups - creates gaps in community leadership and organizing capacity.

Digital Divide: Differences in digital access and skills can exclude some community members from digitally-enabled community building efforts.

Community Building Strategies for Different Contexts
Different types of communities face different challenges and opportunities for building social connections and resilience.

Urban Community Building

Neighborhood-Level Focus: Building community within specific neighborhoods or blocks rather than trying to organize entire cities.

Vertical Community Building: Creating connections among residents of apartment buildings and high-density housing.

Public Space Activation: Making better use of parks, plazas, and other public spaces for community activities and interaction.

Local Business Integration: Working with local businesses to create community gathering spaces and support local economic networks.

Suburban Community Building

Breaking Down Isolation: Creating opportunities for interaction among residents who might otherwise have little contact with neighbors.

Family-Oriented Activities: Organizing activities that bring together families with children while also creating adult social connections.

Shared Resource Development: Organizing tool libraries, community gardens, and other shared resources that create reasons for interaction.

Transportation Solutions: Developing community solutions for transportation challenges that can isolate suburban residents.

Rural Community Building

Regional Cooperation: Building connections among small communities that can share resources and support each other.

Economic Development Focus: Organizing around economic development initiatives that can provide local employment and business opportunities.

Service Delivery Coordination: Collaborating to maintain or improve access to services that might not be economically viable for individual small communities.

Cultural Preservation: Organizing activities that preserve and celebrate rural culture and traditions while adapting to modern challenges.

The Future of Community

As AI continues transforming work and society, communities that proactively strengthen social bonds and build collective capacity will be better positioned to help their members thrive through ongoing changes. The goal isn't to resist technological change but to ensure that human communities remain strong enough to guide how technology is used and to provide the meaning, support, and connection that technology cannot supply.

Strong communities can help individuals navigate AI disruption more successfully while ensuring that technological advancement serves human flourishing rather than fragmenting the social bonds that make life meaningful. In our next chapter, we'll explore how communities and families can prepare the next generation for success in an AI-integrated world, focusing on the skills, values, and capabilities that young people will need to thrive in a rapidly changing technological landscape.

12

PREPARING THE NEXT GENERATION

"We cannot always build the future for our youth, but we can build our youth for the future." – Franklin D. Roosevelt

Twelve-year-old Maya Patel finished her math homework in fifteen minutes using an AI tutoring app that adapted to her learning style, provided instant feedback, and generated practice problems tailored to her specific knowledge gaps. Her essay for English class was polished to perfection after she collaborated with an AI writing assistant that helped her organize her thoughts, improve her grammar, and enhance her vocabulary. For her science project on climate change, an AI research assistant helped her find credible sources, analyze data, and create compelling visualizations.

Maya's academic performance had never been better. Her grades were excellent, her assignments were always completed on time, and her teachers praised her work. But her mother, Dr. Priya Patel, noticed something troubling: Maya seemed to have lost confidence in her own intellectual abilities.

> **"Whenever she faced a problem that required thinking through something step by step, she would immediately reach for her phone to ask an AI," Dr. Patel told me during our interview. "She'd become anxious if she couldn't get instant AI assistance, as if she'd forgotten that her own mind was capable of working through challenges. I started wondering if we were raising a generation of children who would be brilliant with AI support but helpless without it."**

Dr. Patel's concerns reflect a broader challenge facing parents, educators, and communities: How do we prepare young people for success in an AI-integrated world while ensuring they develop the distinctly human capabilities that will remain valuable throughout their lives? This isn't about rejecting AI tools - they're already integral to how children learn and will work - but about raising young people who can use AI effectively while maintaining intellectual independence, creative thinking, and strong human relationships.

The stakes couldn't be higher. The children and teenagers growing up today will spend their entire careers navigating technological change that we can barely imagine. They need skills and capabilities that can adapt to multiple possible futures while preserving what makes them uniquely human.

The New Educational Imperatives

Traditional education was designed for a world where knowledge was scarce and access to information was limited. In an AI world where information is abundant and processing is instant, educational priorities must shift toward capabilities that complement rather than compete with artificial intelligence.

From Knowledge Acquisition to Knowledge Synthesis: Instead of focusing primarily on memorizing facts and procedures, education must emphasize the ability to synthesize information from multiple sources, evaluate credibility, and draw insights that go beyond what AI systems can provide.

From Individual Competence to Collaborative Intelligence:

While individual skills remain important, young people need to learn how to collaborate effectively with both humans and AI systems, understanding when to rely on each and how to integrate different types of intelligence.

From Standardized Solutions to Creative Problem-Solving: AI systems excel at applying known solutions to standard problems. Young people need to develop the creativity and innovative thinking required for novel challenges that don't have predetermined answers.

From Certainty to Ambiguity Tolerance: In a rapidly changing world, the ability to work effectively despite uncertainty, incomplete information, and changing circumstances becomes more valuable than the ability to perform well in predictable situations.

From Passive Learning to Active Creation: Rather than just consuming information and reproducing it on tests, young people need opportunities to create original work, develop their own ideas, and contribute something new to the world.

Sarah Chen, the software developer we've followed throughout this book, has been thinking deeply about these issues as she watches her eight-year-old son navigate school. "I see him getting incredible AI assistance with his schoolwork, but I want to make sure he's still learning to think for himself," she said. "The question isn't whether he should use AI - he definitely should. The question is how to help him use it in ways that make him stronger rather than more dependent."

Core Human Capabilities for an AI World

While we can't predict exactly what future jobs will look like, we can identify core human capabilities that are likely to remain valuable regardless of technological advancement. Developing these capabilities should be central to how we prepare young people for an AI-integrated future.

Critical and Creative Thinking

Independent Analysis: The ability to analyze information, identify assumptions, evaluate evidence, and draw conclusions without relying entirely on AI-generated analysis.

Creative Problem-Solving: Developing novel approaches to complex challenges, thinking outside established patterns, and generating solutions that haven't been programmed into AI systems.

Systems Thinking: Understanding how different parts of complex systems interact, recognizing unintended consequences, and thinking holistically about problems and solutions.

Questioning Skills: Learning to ask good questions, challenge assumptions, and probe deeper into issues rather than accepting initial answers or explanations.

Pattern Recognition Across Domains: Identifying patterns and connections across different fields of knowledge in ways that might not be obvious to AI systems trained on specific datasets.

Emotional and Social Intelligence

Empathy and Perspective-Taking: Understanding others' emotions, motivations, and viewpoints in ways that go beyond algorithmic emotion recognition.

Relationship Building: Developing deep, trusting relationships with others based on mutual understanding, vulnerability, and genuine care.

Conflict Resolution: Navigating disagreements, finding common ground, and helping others resolve disputes through human understanding rather than algorithmic mediation.

Cultural Competence: Understanding and working effectively across different cultural contexts, recognizing how culture shapes thinking and behavior in ways that AI might miss.

Leadership and Influence: Inspiring and motivating others, building consensus, and guiding groups toward shared goals through human connection rather than algorithmic optimization.

Communication and Expression

Authentic Voice Development: Finding and expressing their own unique perspective, style, and voice rather than defaulting to AI-generated content.

Multi-Modal Communication: Effectively communicating through speaking, writing, visual arts, and other modes while understanding how different audiences and contexts require different approaches.

Storytelling and Narrative: Creating compelling stories that connect with human experiences and emotions in ways that purely data-driven content cannot.

Persuasion and Influence: Understanding how to persuade and influence others through authentic human connection rather than algorithmic manipulation.

Cross-Cultural Communication: Communicating effectively with people from different backgrounds, cultures, and generations.

Adaptability and Resilience

Learning How to Learn: Developing meta-cognitive skills that enable rapid acquisition of new knowledge and capabilities as circumstances change.

Uncertainty Tolerance: Working effectively despite incomplete information, changing circumstances, and unpredictable outcomes.

Failure Recovery: Learning from mistakes, bouncing back from setbacks, and maintaining motivation despite temporary failures.

Identity Flexibility: Maintaining core values and sense of self while adapting to changing roles, circumstances, and opportunities.

Stress Management: Developing healthy coping mechanisms for managing the stress and anxiety that accompany rapid change and uncertainty.

Educational Approaches for the AI Age

Developing these capabilities requires educational approaches that go beyond traditional classroom instruction and standardized testing. Young people need learning experiences that are active, collaborative, creative, and connected to real-world challenges.

Project-Based Learning

Real-World Problem Solving: Students work on authentic problems that matter to their communities, requiring them to research, analyze, create, and communicate in ways that mirror real-world work.

Cross-Disciplinary Integration: Projects that require students to integrate knowledge from multiple subjects, developing the systems thinking that complex problems require.

Extended Timeline Projects: Learning experiences that unfold over weeks or months, teaching students to manage complex projects and persist through challenges.

Community Connection: Projects that connect students with community members, businesses, and organizations, providing real audiences for their work and authentic contexts for learning.

Reflection and Iteration: Building in regular opportunities for students to reflect on their learning, get feedback, and improve their work through multiple iterations.

Collaborative Learning

Peer Teaching: Opportunities for students to teach each other, developing communication skills while deepening their own understanding.

Group Problem Solving: Learning experiences that require students to work together to solve complex problems that no individual could handle alone.

Diverse Perspective Integration: Deliberately creating learning groups that include students with different backgrounds, perspectives, and strengths.

Conflict and Consensus Building: Learning experiences that require students to negotiate differences, build consensus, and work together despite disagreements.

Leadership Development: Rotating leadership opportunities so all students develop experience guiding groups and taking responsibility for collective outcomes.

Experiential Learning

Internships and Apprenticeships: Real-world work experiences that allow students to see how their learning applies in professional contexts while building relationships with adult mentors.

Service Learning: Combining academic learning with community service in ways that benefit communities while providing authentic learning contexts for students.

Outdoor and Adventure Learning: Experiences that challenge students physically and mentally while building confidence, teamwork, and resilience.

Cultural Immersion: Opportunities to experience different cultures and communities, developing empathy and global perspective.

Maker Spaces and Hands-On Creation: Learning environments where students can build, create, and experiment with physical materials and tools.

The Role of Technology in Education

The goal isn't to eliminate technology from education but to use it in ways that enhance rather than replace human development. This requires thoughtful integration that leverages AI's strengths while preserving opportunities for students to develop their own capabilities.

AI as Learning Partner, Not Replacement

Collaborative Learning Tools: Using AI to facilitate collaboration among students rather than replacing peer interaction.

Personalized Practice: AI systems that adapt to individual learning needs while ensuring students still grapple with challenging problems independently.

Feedback and Reflection: AI tools that provide feedback on student work while encouraging self-reflection and peer review.

Research Assistance: AI that helps students find and evaluate sources while requiring them to synthesize information and draw their own conclusions.

Creative Inspiration: AI tools that provide creative prompts and suggestions while requiring students to develop and express their own ideas.

Digital Literacy and AI Understanding

How AI Works: Basic understanding of how AI systems process information, make decisions, and learn from data, without requiring technical programming skills.

AI Limitations: Understanding what AI systems can and cannot do, when they're reliable, and when human judgment is necessary.

Ethical AI Use: Learning about bias, privacy, and ethical considerations in AI development and deployment.

Data Literacy: Understanding how to collect, analyze, and interpret data while recognizing the limitations and potential misinterpretations of data-driven insights.

Digital Citizenship: Developing responsible approaches to digital communication, online behavior, and technology use.

Balancing Digital and Analog Experiences

Screen-Free Learning Time: Regular opportunities for learning and creating without digital devices, maintaining the ability to think and work without technological assistance.

Physical and Mental Challenges: Learning experiences that require sustained attention, physical effort, and mental persistence without digital shortcuts.

Face-to-Face Communication: Prioritizing in-person interaction and communication skills alongside digital communication competence.

Handwriting and Mental Math: Maintaining basic capabilities that support cognitive development and provide backup skills when technology isn't available.

Nature and Outdoor Experience: Regular contact with natural environments that provide different types of learning and development opportunities than digital environments.

Parenting in the AI Age

Parents play crucial roles in helping children develop healthy relationships with technology while building the human capabilities that will serve them throughout their lives.

Modeling Healthy Technology Use

Intentional Technology Use: Demonstrating thoughtful choices about when and how to use digital devices rather than constant, unreflective use.

AI Transparency: Explaining when and why you use AI tools while showing children how you maintain your own thinking and decision-making capabilities.

Digital Boundaries: Establishing clear boundaries around technology use for the whole family, including device-free times and spaces.

Critical Evaluation: Modeling critical evaluation of online information, AI-generated content, and digital recommendations.

Human-First Priorities: Demonstrating that human relationships, face-to-face communication, and offline activities remain priorities despite digital conveniences.

Encouraging Independent Thinking

Problem-Solving Before AI: Encouraging children to try solving problems independently before turning to AI assistance.

Question Everything: Teaching children to question information sources, challenge assumptions, and think critically about what they read and hear.

Trial and Error Learning: Allowing children to make mistakes and learn from them rather than preventing all errors through AI assistance.

Boredom and Creativity: Allowing periods of boredom that can spark creativity and independent thinking rather than constantly providing digital entertainment.

Real-World Challenges: Providing children with age-appropriate real-world challenges that require sustained effort and problem-solving.

Building Human Connections

Family Conversations: Regular family discussions about values, current events, and important life questions that require human insight and perspective.

Community Involvement: Involving children in community activities that connect them with people of different ages and backgrounds.

Service to Others: Opportunities for children to help others and contribute to their communities in ways that build empathy and social connection.

Extended Family and Elder Connection: Maintaining connections with grandparents and older adults who can provide historical perspective and life wisdom.

Cultural Traditions: Participating in cultural traditions and practices that connect children to their heritage and community identity.

Maria Santos, the former customer service representative who now runs workshops for displaced workers, has been thoughtful about raising her teenage daughter in the AI age.

> **"I want her to see AI as a powerful tool that can help her do amazing things, but not as a substitute for her own mind and heart," Maria explained. "We have regular conversations about the difference between getting AI help with homework and having AI do the homework for her. I want her to use technology to amplify her own capabilities, not replace them."**

Developing Entrepreneurial Thinking

In an AI-disrupted economy, young people may need to create their own opportunities rather than fitting into traditional career paths.

Developing entrepreneurial thinking - the ability to identify opportunities, take initiative, and create value - becomes increasingly important.

Opportunity Recognition: Learning to identify problems that need solving and unmet needs in their communities or markets.

Initiative and Self-Direction: Developing the ability to start projects and pursue goals without external direction or supervision.

Resource Management: Understanding how to identify, acquire, and manage resources needed to accomplish goals.

Risk Assessment and Management: Learning to evaluate risks and take calculated chances rather than avoiding all uncertainty.

Value Creation: Understanding how to create value for others through products, services, or solutions that meet real needs.

Persistence and Resilience: Developing the ability to persist through setbacks and failures while learning from mistakes.

Network Building: Learning to build relationships and networks that can provide support, opportunities, and collaboration.

Ethical Development in an AI World

As AI systems become more powerful and pervasive, young people need strong ethical foundations to guide their decisions about how to use these tools responsibly.

Human Dignity and Worth: Understanding that human beings have inherent worth and dignity that should guide how technology is developed and used.

Privacy and Consent: Learning about privacy rights and the importance of informed consent in data collection and use.

Fairness and Justice: Understanding how AI systems can perpetuate or amplify bias and discrimination, and developing commitment to fairness and justice.

Transparency and Accountability: Valuing transparency in AI systems and holding developers and users accountable for the impacts of their choices.

Environmental Responsibility: Understanding the environmental costs of digital technology and making responsible choices about technology use.

Global Perspective: Recognizing that AI development and deployment affect people around the world, often in different ways.

Democratic Values: Understanding how AI affects democratic institutions and citizen participation in governance.

Career Exploration and Development

While we can't predict specific future jobs, we can help young people develop frameworks for exploring careers and adapting to changing opportunities throughout their working lives.

Values-Based Career Exploration: Helping young people identify their core values and interests rather than just focusing on job titles or salary expectations.

Skill Portfolio Development: Encouraging development of diverse skills that can be combined in different ways rather than narrow specialization in single domains.

Industry and Sector Awareness: Understanding how different industries and sectors are being affected by AI while identifying areas of opportunity and growth.

Networking and Relationship Building: Learning to build professional relationships and networks that can provide information, opportunities, and support throughout their careers.

Continuous Learning Mindset: Developing attitudes and habits that support lifelong learning and adaptation to changing circumstances.

Entrepreneurial Options: Understanding entrepreneurship and self-employment as viable career options rather than just traditional employment.

Global Perspective: Recognizing that career opportunities may be global rather than just local, and developing skills for working across cultures and time zones.

Supporting Diverse Learning Styles and Needs

Young people have different learning styles, interests, and capabilities. Preparing them for an AI world means recognizing and supporting this diversity rather than trying to fit everyone into the same educational approach.

Multiple Intelligence Recognition: Recognizing that people have different types of intelligence - linguistic, logical-mathematical, spatial, musical, kinesthetic, interpersonal, intrapersonal, and naturalist - and providing opportunities for all types to be developed and valued.

Individual Learning Plans: Developing educational approaches that can be customized to individual interests, learning styles, and goals rather than one-size-fits-all curricula.

Strength-Based Development: Focusing on developing individual strengths while providing support for areas of challenge rather than trying to make everyone equally capable in all areas.

Cultural Responsiveness: Educational approaches that recognize and value different cultural backgrounds, learning traditions, and ways of knowing.

Special Needs Support: Ensuring that students with disabilities, learning differences, or other special needs receive appropriate support while developing their unique capabilities.

Economic Accessibility: Making high-quality educational opportunities available to young people regardless of their family's economic circumstances.

The Community's Role

Preparing young people for an AI-integrated future isn't just the responsibility of parents and schools - it requires entire communities working together to create environments that support youth development.

Mentorship Programs: Connecting young people with adult mentors who can provide guidance, support, and real-world perspective on career and life decisions.

Community Service Opportunities: Providing opportunities for young people to contribute to their communities while developing skills and building relationships with people of different ages.

Business and Organization Partnerships: Creating partnerships between schools and local businesses, nonprofits, and organizations that can provide authentic learning experiences and career exploration opportunities.

Cultural and Arts Programs: Supporting programs that help young people develop creative expression, cultural awareness, and artistic capabilities that AI cannot replicate.

Sports and Recreation: Maintaining programs that build physical fitness, teamwork, leadership, and resilience through athletic and recreational activities.

Faith and Values Communities: Supporting religious and spiritual communities that can provide young people with ethical foundations, meaning, and purpose beyond material success.

Measuring Success in New Ways

As we reimagine education for an AI world, we also need to develop new ways of measuring success that go beyond standardized test scores and traditional academic metrics.

Portfolio-Based Assessment: Evaluating student learning through portfolios of work that demonstrate growth, creativity, and real-world application of knowledge and skills.

Community Impact Measures: Assessing how well students can contribute to their communities and solve real-world problems.

Collaboration and Leadership Skills: Measuring students' ability to work effectively with others and provide leadership in group settings.

Adaptability and Resilience: Assessing how well students handle change, recover from setbacks, and continue learning in new situations.

Creative and Original Thinking: Evaluating students' ability to generate novel ideas and approaches rather than just reproduce existing knowledge.

Ethical Reasoning: Assessing students' ability to think through ethical dilemmas and make decisions based on sound moral principles.

Life Satisfaction and Well-Being: Measuring whether educational experiences contribute to students' overall life satisfaction, mental health, and sense of purpose.

The goal isn't to protect young people from AI or to prepare them for a world without technology. Instead, the goal is to help them develop the distinctly human capabilities that will allow them to thrive alongside AI while maintaining their humanity, creativity, and sense of purpose. Young people who grow up with strong critical thinking skills, deep empathy, creative problem-solving abilities, and ethical foundations will be equipped not just to survive in an AI world but to help shape it in ways that serve human flourishing.

In our final chapter, we'll explore what the future of human work might look like as AI capabilities continue advancing, and how we can create economic systems that value and reward uniquely human contributions while leveraging AI's capabilities for the benefit of all.

13

THE FUTURE OF HUMAN WORK

"The future belongs to those who learn more skills and combine them in creative ways." – Robert Greene

In 2030, Dr. Amara Okafor starts her workday not in a traditional office but in what she calls her "collaboration pod" - a home workspace designed for seamless interaction with both AI systems and human colleagues around the world. As a "Relationship Intelligence Specialist," her job didn't exist when she graduated from college in 2021, but it represents exactly the kind of work that has emerged as AI has transformed the global economy.

Dr. Okafor's morning begins with reviewing insights generated overnight by AI systems that analyzed customer interaction patterns, market sentiment, and relationship health metrics across the dozens of client partnerships her team manages. But the AI doesn't tell her what to do with this information - instead, it provides the data foundation for decisions that require distinctly human judgment about trust, cultural nuance, and long-term relationship building. Her first client meeting involves a complex three-way partnership between a renewable energy company in Denmark, a battery manufacturer in South Korea, and a grid management cooperative in Kenya. The AI systems have identified potential friction points in their collaboration and suggested multiple strategic options. But Dr. Okafor's role is to understand the human dynamics, cultural contexts, and relationship history that will determine which strategy actually works.

"The AI can tell us what the data suggests," she explains to the client team during their video conference, "but only humans can understand what the people behind that data actually care about, what they're afraid of, and what will motivate them to work together successfully over the long term." After facilitating a breakthrough in the partnership strategy, Dr. Okafor spends her afternoon mentoring junior colleagues across three continents, reviewing AI-generated content for emotional authenticity, and designing a new framework for cross-cultural trust building that combines AI analysis with human insight.

Her work is intellectually challenging, financially rewarding, and deeply meaningful. It requires capabilities that no AI system can replicate - empathy, cultural intelligence, creative problem-solving, and the ability to build genuine human connections. Yet it would be impossible without AI systems that handle data processing, pattern recognition, and routine analysis. Dr. Okafor's career illustrates what the future of human work might look like: not humans competing with AI, but humans and AI collaborating to create value that neither could produce alone. This future isn't guaranteed - it will require deliberate choices about how we design AI systems, structure economies, and value human contributions. But it represents one of the most promising possibilities for maintaining meaningful, well-compensated human work in an AI-integrated world.

The Transformation of Work Categories

As AI capabilities continue expanding, the nature of work itself is being redefined in ways that go beyond simple job displacement.

Understanding these transformations helps us envision how human work might evolve over the coming decades.

From Tasks to Outcomes

Traditional jobs were organized around specific tasks - data entry, report writing, customer service calls, inventory management. AI excels at many individual tasks, but humans remain superior at managing complex outcomes that require integrating multiple considerations, adapting to unexpected circumstances, and balancing competing priorities.

Outcome-Based Roles: Future human work increasingly focuses on achieving specific outcomes rather than completing predetermined tasks. A marketing professional might be responsible for "building brand trust in specific communities" rather than "writing social media posts" or "analyzing engagement metrics."

Systems Integration: Humans excel at integrating outputs from multiple AI systems, understanding how different pieces of information relate to broader goals, and making decisions when AI recommendations conflict or prove inadequate.

Context and Judgment: While AI can process information faster than humans, human workers provide the contextual understanding and judgment necessary to interpret what that information means for specific situations, relationships, and goals.

Adaptive Problem-Solving: When standard procedures fail or unexpected challenges arise, human workers can improvise solutions, think creatively, and adapt approaches in ways that rigid AI systems cannot.

From Individual to Collaborative Intelligence

The future of work increasingly involves collaboration between humans and AI systems, with humans focusing on areas where their capabilities complement rather than compete with artificial intelligence.

Human-AI Teams: Work teams include both human members and AI systems, with humans handling relationship management, creative thinking, and ethical judgment while AI handles data processing, pattern recognition, and routine analysis.

AI Direction and Oversight: Humans increasingly work as AI managers, directing AI systems toward specific goals, evaluating AI outputs for quality and appropriateness, and taking responsibility for decisions based on AI analysis.

Cross-Domain Integration: Humans excel at connecting insights across different domains and industries in ways that specialized AI systems might miss, creating opportunities for innovation and creative problem-solving.

Emotional and Cultural Intelligence: As global collaboration increases, human skills in understanding cultural differences, managing emotional dynamics, and building trust across diverse groups become more valuable.

From Expertise to Meta-Skills

Traditional career success often depended on deep expertise in specific technical areas. Future success may depend more on meta-skills - capabilities that help people learn, adapt, and collaborate effectively across changing circumstances.

Learning and Adaptation: The ability to rapidly acquire new knowledge, adapt to changing technologies, and integrate insights from multiple sources becomes more valuable than fixed expertise in specific areas.

Communication and Translation: Skills in translating between different audiences, explaining complex concepts to non-experts, and facilitating understanding across diverse groups become increasingly important.

Relationship Building and Maintenance: The ability to build trust, maintain long-term relationships, and navigate complex interpersonal dynamics remains distinctly human and increasingly valuable.

Creative Problem-Solving: When standard solutions don't work, humans provide the creative thinking needed to develop novel approaches to complex challenges.

Ethical Reasoning: As AI systems make more decisions, humans provide the ethical judgment and moral reasoning needed to ensure that technological capabilities serve human values and well-being.

Emerging Categories of Human Work

As AI transforms traditional job categories, new types of work are emerging that specifically leverage human capabilities while benefiting from AI assistance.

Human-AI Collaboration Specialists

These roles focus on optimizing the interaction between human intelligence and AI systems to achieve better outcomes than either could produce independently.

AI Training and Fine-Tuning: Specialists who help train AI systems to work effectively in specific contexts, ensuring that AI outputs align with human values and organizational goals.

Human-AI Interface Design: Professionals who design systems and processes that enable effective collaboration between humans and AI, ensuring that human workers can leverage AI capabilities without losing their own skills and judgment.

AI Ethics and Oversight: Specialists who monitor AI systems for bias, unintended consequences, and ethical concerns while developing policies and procedures for responsible AI use.

Performance Integration Analysis: Professionals who analyze how well human-AI teams are working together and optimize these collaborations for better outcomes.

Relationship and Trust Architects

As digital interactions increase, specialists in building and maintaining human relationships become more valuable.

Community Builders: Professionals who design and facilitate communities - both online and offline - that bring people together around shared interests, challenges, or goals.

Trust and Reputation Managers: Specialists who help individuals and organizations build and maintain trust in an increasingly digital world where authentic relationships are harder to establish.

Cultural Bridge-Builders: Professionals who help people from different cultural backgrounds understand each other and work together effectively, especially important as remote work enables more global collaboration.

Conflict Resolution and Mediation: Specialists who help resolve disputes and facilitate understanding between parties in conflict, skills that remain distinctly human despite technological advancement.

Experience and Meaning Creators
As AI handles more routine tasks, humans focus increasingly on

creating experiences that provide meaning, beauty, and emotional fulfillment.

Experience Designers: Professionals who create meaningful experiences - educational, recreational, spiritual, or artistic - that enrich human life in ways that purely digital experiences cannot.

Storytelling and Narrative Specialists: Professionals who help individuals and organizations craft authentic stories that connect with human experiences and emotions.

Wellness and Life Coaching: Specialists who help people navigate life transitions, develop personal meaning and purpose, and build resilience for handling change and uncertainty.

Creative Directors and Curators: Professionals who guide creative projects, curate artistic experiences, and help distinguish between AI-generated and human-created content based on aesthetic and emotional criteria.

Complexity and Systems Navigators

As systems become more complex and interconnected, specialists who can understand and navigate complexity become increasingly valuable.

Systems Thinking Consultants: Professionals who help organizations and communities understand complex interactions, unintended consequences, and systemic solutions to multi-faceted problems.

Integration Specialists: Professionals who help organizations integrate multiple AI systems, technologies, and processes while ensuring that human needs and values remain central.

Future Scenario Planners: Specialists who help organizations and communities prepare for multiple possible futures, developing adaptive strategies that work across different scenarios.

Complexity Translators: Professionals who help non-experts understand complex systems and technologies, facilitating informed decision-making about technological adoption and policy.

Care and Development Facilitators

Human-centered services that require empathy, understanding, and personal connection remain distinctly valuable and difficult to automate.

Personalized Learning Facilitators: Educators who design and facilitate learning experiences tailored to individual needs, interests, and goals rather than standardized curricula.

Life Transition Coaches: Specialists who help people navigate major life changes - career transitions, relationship changes, health challenges, or family developments.

Elder Care and Support Specialists: Professionals who provide companionship, advocacy, and support for older adults, combining practical assistance with genuine human connection.

Child Development Specialists: Professionals who support families and communities in raising children with strong human capabilities alongside technological fluency.

New Economic Models for Human Work

The future of human work may require new economic models that better recognize and reward uniquely human contributions while accounting for AI's role in value creation.

Stakeholder Capitalism Evolution

Traditional capitalism focuses primarily on shareholder returns, but future economic models may give greater weight to all stakeholders - employees, communities, customers, and society as a whole.

Shared Value Creation: Business models that create value for multiple stakeholders simultaneously rather than maximizing returns for shareholders at the expense of others.

Community Wealth Building: Economic approaches that keep wealth circulating within communities through local ownership, cooperative businesses, and community investment.

Social Impact Measurement: Accounting systems that measure and reward social and environmental impacts alongside financial returns.

Stakeholder Governance: Corporate governance systems that include employee representatives, community members, and other stakeholders in major decision-making processes.

Cooperative and Platform Economics

New forms of economic organization may provide workers with more control over AI systems and greater shares of the value they help create.

Worker Cooperatives: Businesses owned and controlled by their workers, ensuring that productivity gains from AI systems benefit those who work alongside them.

Platform Cooperatives: Digital platforms owned and governed by their users rather than external shareholders, providing workers with more control over how AI systems are developed and deployed.

Community Ownership Models: Local communities owning and controlling AI systems and digital infrastructure, ensuring that technological benefits serve community needs.

Profit-Sharing and Ownership: Systems that provide workers with ownership stakes in the businesses where they work, ensuring they benefit from AI-driven productivity improvements.

Care Economy Recognition

Many forms of valuable human work - caregiving, community building, environmental stewardship - are currently undervalued in traditional economic systems. Future models may better recognize and compensate these contributions.

Care Income Systems: Economic policies that provide income support for care work - raising children, caring for elders, supporting community members - that is essential for social well-being but not captured in traditional markets.

Community Contribution Valuation: Systems for recognizing and compensating community volunteer work, civic participation, and other forms of social contribution.

Environmental Stewardship Compensation: Payment systems for environmental restoration, conservation, and stewardship work that provides public benefits.

Cultural Preservation Support: Economic support for maintaining cultural traditions, artistic practices, and community knowledge that enriches human life but doesn't generate traditional market returns.

The Geography of Future Work

AI is also transforming where work happens, creating new possibilities for geographic distribution of economic opportunity while also creating new challenges for community development.

Distributed Work Networks

AI-enabled communication and collaboration tools make it increasingly possible for people to work together effectively regardless of geographic location.

Global Talent Access: Organizations can access talent from anywhere in the world, potentially creating opportunities for people in economically disadvantaged regions while also creating new forms of competition.

Rural Renaissance: High-quality internet access and AI tools may enable rural areas to participate in knowledge work that was previously concentrated in urban areas.

Community-Anchored Remote Work: People may be able to live in communities they choose based on quality of life, family connections, or personal preferences rather than employment opportunities.

Regional Specialization: Different regions may develop specializations in particular types of human-AI collaboration, creating economic clusters based on human capabilities rather than just natural resources or infrastructure.

Local Service Renaissance

While some work becomes globally distributed, other forms of human work become more locally focused as people value authentic, personal experiences and community connections.

Hyperlocal Services: Services that require physical presence and local knowledge - from childcare and elder care to home maintenance and community organizing - may become more valuable as other work becomes digitized.

Artisan and Craft Revival: As mass production becomes increasingly automated, handmade and locally-produced goods may become more valued for their authenticity and human touch.

Community Experience Economy: Local experiences - restaurants, entertainment, recreation, education - that bring people together in physical spaces may become more important as digital interactions increase.

Place-Based Problem Solving: Communities may develop distinctive approaches to local challenges that leverage both AI capabilities and local human knowledge and relationships.

Skills and Capabilities for Future Work

Success in future work environments will require developing capabilities that complement AI while remaining distinctly human.

Meta-Learning and Adaptation

Learning How to Learn: The ability to rapidly acquire new knowledge and skills as circumstances change becomes more valuable than fixed expertise in specific areas.

Pattern Recognition Across Domains: The ability to see connections and patterns across different fields and apply insights from one area to challenges in another.

Uncertainty Navigation: Skills in making good decisions despite incomplete information and changing circumstances.

Feedback Integration: The ability to learn from mistakes, incorporate feedback, and continuously improve performance over time.

Creative and Systems Thinking

Creative Problem-Solving: The ability to generate novel solutions to complex problems that don't have predetermined answers.

Systems Understanding: The ability to understand how different parts of complex systems interact and identify leverage points for creating positive change.

Design Thinking: Approaches to problem-solving that focus on human needs and experiences while integrating technological capabilities.

Innovation and Entrepreneurship: The ability to identify opportunities, create new value, and build sustainable ventures or initiatives.

Social and Emotional Intelligence

Empathy and Perspective-Taking: The ability to understand others' emotions, motivations, and viewpoints in authentic and helpful ways.

Communication and Influence: Skills in persuading, inspiring, and motivating others through authentic human connection rather than algorithmic manipulation.

Conflict Resolution: The ability to help others resolve disagreements and find common ground despite differences.

Cultural Competence: Understanding how to work effectively across different cultural contexts and help bridge cultural differences.

Leadership and Collaboration

Facilitative Leadership: The ability to help groups accomplish their goals by facilitating participation, building consensus, and managing group dynamics.

Network Building: Skills in creating and maintaining professional and personal relationships that provide mutual support and collaboration opportunities.

Team Integration: The ability to work effectively in diverse teams that include both humans and AI systems.

Mentorship and Development: Skills in helping others develop their capabilities and navigate career transitions and challenges.

Potential Challenges and Risks

While the future of human work holds considerable promise, several challenges could prevent these positive outcomes from materializing.

Economic Inequality

AI Access Inequality: If access to advanced AI tools is limited to wealthy individuals and organizations, it could exacerbate existing economic inequalities rather than creating broadly shared benefits.

Skill Development Inequality: People from disadvantaged backgrounds may have less access to the education and training needed to develop the human capabilities that remain valuable in an AI world.

Geographic Inequality: Rural areas and developing regions may lack the infrastructure and resources needed to participate in AI-enhanced economic opportunities.

Intergenerational Inequality: Older workers may face greater challenges adapting to AI-integrated work environments, potentially creating age-based economic disadvantages.

Work Displacement Timeline

Transition Speed Mismatch: AI capabilities may advance faster than people's ability to develop complementary skills, creating periods of significant unemployment and economic disruption.

Retraining Challenges: Current adult education and retraining systems may be inadequate for the scale and speed of change that AI disruption requires.

Community Disruption: Entire communities may lose their economic base faster than they can develop new industries and opportunities.

Social Support Inadequacy: Existing social safety nets may be insufficient to support people through extended periods of economic transition.

Power Concentration

Technology Control: If AI systems remain controlled by a small number of large corporations or wealthy individuals, they may be used primarily to benefit those groups rather than society as a whole.

Market Concentration: AI advantages may enable large organizations to dominate markets and eliminate smaller competitors, reducing economic opportunities for individual workers and small businesses.

Democratic Governance: Concentrated AI power could undermine democratic governance and citizen participation in decisions about how AI systems are developed and deployed.

Labor Organization Challenges: Traditional labor unions and worker organizations may struggle to represent workers in AI-integrated economies, potentially reducing worker bargaining power.

Preparing for Multiple Scenarios

Given the uncertainty surrounding AI development and its economic impacts, individuals and communities should prepare for multiple possible futures rather than betting on any single scenario.

Skill Portfolio Diversification

Human-AI Collaboration: Develop skills in working effectively with AI systems while maintaining distinctly human capabilities.

Domain Flexibility: Cultivate capabilities that can be applied across multiple industries and contexts rather than specializing narrowly.

Local and Global Competence: Develop both local knowledge and relationships and global communication and collaboration skills.

Technical and Human Balance: Maintain both technological fluency and strong human relationship skills.

Economic Resilience Building

Multiple Income Streams: Develop diverse sources of income that aren't all vulnerable to the same types of AI disruption.

Community Economic Engagement: Participate in local economies and community-based economic activities that provide resilience during broader economic changes.

Ownership and Investment: Build ownership stakes in businesses, real estate, or other assets that can provide income independent of employment.

Cooperative and Mutual Aid: Participate in cooperative economic ventures and mutual aid networks that provide support during transitions.

Community and Network Development

Local Community Investment: Build strong relationships and capabilities within local communities that can provide support and opportunities regardless of broader economic changes.

Professional Network Maintenance: Maintain diverse professional networks that can provide information, opportunities, and collaboration across different industries and regions.

Intergenerational Connections: Build relationships across different age groups to access diverse perspectives, knowledge, and support systems.

Global and Cultural Networks: Develop relationships with people from different cultural backgrounds and geographic regions to access diverse opportunities and perspectives.

The Role of Human Choice

Perhaps most importantly, the future of human work will be shaped by the choices we make - as individuals, communities, organizations, and societies - about how we want to integrate AI into our economic systems.

Values-Based Development: We can choose to develop and deploy AI systems in ways that support human flourishing, community well-being, and shared prosperity rather than just maximizing efficiency or profit.

Inclusive Innovation: We can choose to ensure that the benefits of AI development are shared broadly rather than concentrated among a small group of technology developers and capital owners.

Democratic Governance: We can choose to maintain democratic control over AI development and deployment, ensuring that these powerful technologies serve public interests rather than just private ones.

Human-Centered Design: We can choose to design work systems that enhance human capabilities and provide meaningful opportunities for contribution rather than just replacing human workers with automated systems.

Community Resilience: We can choose to strengthen communities and social bonds that provide meaning and support beyond economic relationships.

The future of human work in an AI world is not predetermined. While technological capabilities create constraints and opportunities, human choices about values, policies, and social organization will ultimately determine whether AI serves as a tool for shared prosperity or a source of increased inequality and social disruption. The people whose stories we've followed throughout this book - Marcus Chen finding new purpose in financial education, Maria Santos building community support networks, Jennifer Walsh developing deeper client relationships, Robert Chen discovering his identity beyond his job title - represent the kind of human adaptation and resilience that can help shape a positive future for human work. Their experiences suggest that while AI may eliminate some forms of work, it also creates opportunities for more meaningful, creative, and relationally rich forms of human contribution.

The challenge is ensuring that these opportunities are available to everyone, not just those who start with economic and educational advantages. The future of human work will ultimately depend on our collective commitment to creating economic systems that value and reward the irreplaceable contributions that only conscious, creative, caring human beings can provide. In an age of artificial intelligence, our most important intelligence may be the wisdom to remain fully, authentically, and purposefully human.

As we conclude this exploration of surviving and thriving in an AI-dominated world, the message is clear: the future belongs not to those who can compete with machines, but to those who can collaborate with them while preserving and developing the capabilities that make us uniquely human. The path forward requires both individual adaptation and collective action to ensure that artificial intelligence serves human flourishing rather than replacing it. In our conclusion, we'll synthesize the key insights from our journey and provide a framework for moving forward with both confidence and compassion in the age of AI.

CONCLUSION : HUMANITY'S PLACE IN THE AGE OF MACHINES

"The best time to plant a tree was 20 years ago. The second best time is now." – Chinese Proverb

Five years after losing her job to AI automation, Maria Santos stood before a packed auditorium at Portland Community College, preparing to deliver the keynote address at a conference called "Thriving Together: Human-AI Collaboration in the 21st Century." The audience included displaced workers, community organizers, business leaders, and policymakers - all grappling with questions about how to navigate the ongoing transformation of work and society.

> **"Five years ago, I thought my life was over when a chatbot replaced me," Maria began, her voice steady and confident. "I was terrified that I had become obsolete, that there was no place for someone like me in a world of artificial intelligence. Tonight, I want to tell you that I was both completely wrong and absolutely right."**

She paused, scanning the faces in the audience - faces that reflected the same mixture of fear, hope, and determination she had felt during those difficult early months of transition.

> **"I was wrong because there is a place for all of us in an AI world. But I was right that it's not the same place we occupied before. The question isn't whether AI will change our lives - it already has. The question is whether we'll have a voice in determining what those changes look like and whether they serve human flourishing or just technological advancement."**

Maria went on to describe her journey from displaced customer service representative to community organizer, from someone who felt replaced by technology to someone who helps others find their own paths alongside AI. Her story had become one of transformation rather than loss, of discovering capabilities she never knew she possessed and finding purpose more meaningful than any she had experienced in traditional employment.

As I listened to Maria's keynote, I was struck by how much her perspective had evolved since our first interview years earlier. She had moved through the stages of AI grief we explored in Chapter 3, developed the psychological resilience discussed in Chapter 4, and built the kind of portfolio identity outlined in Chapter 8. Most importantly, she had become living proof that adaptation to AI disruption isn't just about individual survival - it's about collective transformation toward more human-centered ways of living and working together.

The Journey We've Taken Together

Throughout this book, we've traveled alongside people like Maria, Marcus, Jennifer, Robert, Sarah, and others as they've navigated the challenges and opportunities of living in an AI-integrated world. Their stories illustrate that while AI disruption creates real difficulties, it also opens possibilities for more meaningful, creative, and connected ways of being human.

We began in **Part I: The Reckoning** by acknowledging the full scope of the challenge. We explored how AI creates invisible casualties - people whose contributions are devalued or eliminated without recognition or support. We examined how the digital divide is widening, creating new forms of inequality based on access to AI tools and capabilities. We delved into the psychology of adaptation, recognizing that adjusting to AI disruption requires more than just learning new skills - it demands fundamental shifts in how we understand ourselves and our place in the world.

We then addressed the anxiety that accompanies such profound change, offering practical strategies for managing the fear and uncertainty that naturally arise when traditional sources of security and identity are threatened. These chapters established that AI disruption is not just a technological or economic challenge - it's a deeply human challenge that requires psychological, social, and spiritual responses.

Part II: The Response shifted focus to practical adaptation strategies. We explored how people can adjust to AI integration without becoming technology experts, emphasizing that success lies not in competing with machines but in developing distinctly human capabilities that complement artificial intelligence. We examined both the dark side of AI society - the risks of dependency, fragmentation, and loss of human agency - and the bright side - the irreplaceable human skills that become more valuable as AI handles routine tasks.

We investigated how to build new identities that aren't dependent on traditional employment categories and developed financial strategies for surviving and thriving during extended career transitions. These chapters provided concrete tools and frameworks that people can use immediately while building long-term resilience for ongoing change.

Part III: The Rebuild expanded our focus from individual adaptation to collective transformation. We examined what responsible AI adoption looks like from corporate and government perspectives, recognizing that individual efforts alone are insufficient for managing change of this magnitude. We explored how communities can strengthen social bonds and build resilience in an increasingly digital world, and how we can prepare the next generation for success in an AI-integrated future.

Finally, we looked ahead to envision what the future of human work might look like, recognizing that while we cannot predict specific outcomes, we can prepare for multiple scenarios while working to shape developments in directions that serve human flourishing.

Key Insights and Principles

Several fundamental insights have emerged from our exploration that can guide anyone navigating AI disruption:

AI Integration Is Inevitable, But Its Impact Is Not

The development and deployment of AI systems will continue regardless of individual or even collective resistance. The technology provides too many benefits - increased efficiency, enhanced capabilities, cost reduction - for its adoption to be stopped. However, how AI is integrated into our lives and communities remains a matter of choice. We can choose approaches that enhance human capabilities rather than simply replacing human workers, that strengthen rather than fragment communities, and that serve broadly shared prosperity rather than concentrated wealth.

Human Value Lies in Qualities AI Cannot Replicate

Throughout our exploration, we've seen that the humans who thrive alongside AI are those who focus on developing capabilities that artificial intelligence cannot match: authentic empathy, creative problem-solving, ethical reasoning, relationship building, cultural intelligence, and the integration of insights across multiple domains. These aren't consolation prizes or leftover scraps from the AI feast - they're fundamental human capabilities that become more valuable, not less, as AI handles routine cognitive tasks.

Adaptation Requires Both Individual and Collective Action

While personal resilience, skill development, and psychological adaptation are essential, they're insufficient on their own. The scale and speed of AI disruption require coordinated responses from governments, businesses, educational institutions, and communities. The most successful adaptation efforts we've observed combine individual initiative with collective support, personal growth with policy change, and private adaptation with public investment in systems that help everyone navigate technological transformation.

Community and Connection Are More Important Than Ever

As AI integration proceeds, the quality of human relationships and community bonds becomes increasingly crucial for both practical and psychological well-being. Strong communities provide mutual aid during transitions, networks for discovering opportunities, emotional support during difficult changes, and collective power to influence how AI is developed and deployed. Investment in community building isn't a nice-to-have addition to AI adaptation - it's a fundamental requirement for individual and collective thriving.

The Future Depends on Choices We Make Now

The future of human work and society in an AI world is not predetermined by technological capabilities. It will be shaped by the choices we make - as individuals, families, communities, organizations, and societies - about how we want to integrate AI into our lives. We can choose approaches that enhance human dignity and expand opportunities for meaningful contribution, or we can drift toward outcomes that primarily serve technological efficiency and capital accumulation. The window for making these choices deliberately and democratically is open now, but it won't remain open indefinitely.

A Framework for Moving Forward

Based on the experiences and insights we've explored throughout this book, here's a practical framework that anyone can use to navigate AI integration while maintaining their humanity and building resilience for ongoing change:

1. Develop Self-Knowledge and Values Clarity

Begin by identifying your core values, character strengths, and sources of meaning that exist independently of your job title or economic function. Use the identity reconstruction approaches from Chapter 8 to build a sense of self that can adapt to changing circumstances while remaining anchored in what matters most to you. This foundation provides stability during transitions and guidance for making decisions about how to engage with AI tools and opportunities.

2. Cultivate Distinctly Human Capabilities

Deliberately develop the capabilities that AI cannot replicate: empathy, creativity, ethical reasoning, relationship building, cultural intelligence, and systems thinking. These aren't just career insurance - they're the foundation for a fulfilling human life regardless of technological developments. Use the strategies from Chapter 7 to identify and strengthen your unique human contributions.

3. Learn to Collaborate with AI

Rather than avoiding or competing with AI systems, learn to work alongside them effectively. Develop skills in directing AI tools, evaluating AI outputs, and integrating AI capabilities with human judgment. Use the approaches from Chapter 5 to become proficient with AI assistance without losing your own cognitive capabilities or decision-making independence.

4. Build Financial Resilience

Create financial strategies that account for uncertainty and change rather than assuming stable employment and predictable income growth. Develop multiple income streams, build extended emergency funds, invest in skills and capabilities that retain value across changing circumstances, and participate in community-based economic activities that provide both financial benefits and social connection.

5. Strengthen Community Connections

Invest deliberately in building and maintaining relationships within your local community and professional networks. Participate in mutual aid, skill sharing, and collective problem-solving activities that strengthen social bonds while providing practical benefits. Use both digital tools and face-to-face interactions to create communities that can support members through transitions and changes.

6. Engage in Collective Action

While individual adaptation is essential, systemic challenges require collective responses. Participate in efforts to advocate for responsible AI development, support policies that help communities navigate technological transitions, and create institutions that serve human needs rather than just technological capabilities. This might involve political engagement, community organizing, cooperative business development, or supporting organizations that work on AI ethics and policy.

---

7. Maintain Hope and Agency

Perhaps most importantly, maintain belief in human agency and the possibility of positive outcomes despite uncertainty and challenges. The people whose stories we've followed throughout this book succeeded not because they had perfect information or made perfect choices, but because they maintained hope and continued taking constructive action even when outcomes were uncertain.

The Continuing Story

The stories we've followed throughout this book don't end with this conclusion - they continue evolving as AI capabilities advance and society adapts to ongoing technological change. Maria Santos continues organizing communities and advocating for worker protections. Marcus Chen has expanded his financial education work and is exploring cooperative business models. Jennifer Walsh has helped launch a community investment fund that supports local businesses. Robert Chen is mentoring other people through career transitions while developing his own consulting practice.

Their ongoing experiences remind us that adaptation to AI disruption isn't a problem to be solved once and forgotten - it's an ongoing process of growth, learning, and community building that will continue throughout their lives and careers. The strategies and insights they've developed provide foundations for continued adaptation rather than final answers to AI challenges.

The Larger Stakes

While this book has focused primarily on practical strategies for individual and community adaptation, it's important to recognize the larger stakes involved in how humanity navigates the AI revolution. The choices we make in the coming years about AI development and deployment will shape society for generations to come. If we allow AI development to proceed without deliberate attention to human values and community well-being, we risk creating a world where technological efficiency is valued over human dignity, where economic benefits are concentrated among technology owners while costs are borne by everyone else, and where human capabilities atrophy from lack of use and value. But if we make conscious choices to develop and deploy AI in ways that enhance rather than replace human capabilities, that strengthen rather than fragment communities, and that serve broadly shared prosperity rather than concentrated wealth, we have the opportunity to create a world where technology truly serves human flourishing.

The window for making these choices deliberately and democratically is open now. The people whose voices are included in these decisions - and the values that guide those decisions - will determine whether AI becomes a tool for human liberation or a source of increased inequality and social control.

A Personal Invitation

As we conclude this journey together, I want to extend a personal invitation to each reader: don't just consume the information in this book - act on it. The strategies and insights we've explored only create value when they're applied in real life, adapted to specific circumstances, and shared with others facing similar challenges. Start with small steps. Identify one human capability you want to develop more fully. Reach out to one neighbor you don't know well. Learn to use one AI tool in a way that enhances rather than replaces your own thinking. Join one community organization that's working on issues you care about. Have one conversation with a family member about how you can support each other through technological changes. These small actions, multiplied across millions of people, constitute the collective adaptation that will determine whether AI serves human thriving or merely technological advancement.

The Enduring Human Questions

As artificial intelligence capabilities continue expanding, we're compelled to grapple with fundamental questions about what makes life meaningful, what kinds of communities we want to live in, and what we value most deeply about human existence. These are not new questions - humans have always had to decide what kinds of lives are worth living and what kinds of societies are worth building. But AI forces us to address these questions with new urgency and clarity because technological capabilities are advancing faster than our wisdom about how to use them well.

The answers we develop - individually and collectively - to questions about human purpose, community connection, economic justice, and technological governance will shape not just our own lives but the kind of world we leave for future generations.

Looking Ahead with Courage and Compassion

The age of artificial intelligence presents both unprecedented challenges and unprecedented opportunities for human flourishing. We cannot know exactly what the future holds, but we can face that uncertainty with courage, compassion, and commitment to the values and relationships that make life meaningful. The people whose stories we've shared throughout this book demonstrate that it's possible to adapt successfully to AI disruption while maintaining human dignity, building stronger communities, and finding deeper sources of meaning and purpose.

Their experiences provide hope and practical guidance for anyone grappling with similar challenges. But their stories also remind us that successful adaptation requires both individual initiative and collective support, both practical skill development and psychological resilience, both technological fluency and commitment to essentially human values.

As we move forward into an uncertain but possibility-rich future, we can take inspiration from the resilience, creativity, and community-building capacity that humans have demonstrated throughout history when facing major technological and social transitions. We can trust in our ability to learn, adapt, and create meaning even in changing circumstances.

Most importantly, we can remember that in an age of artificial intelligence, our greatest intelligence may be the wisdom to remain fully, authentically, and purposefully human - not in opposition to technology, but in thoughtful integration with it in service of the values and relationships that make life worth living.

The future is not something that happens to us - it's something we create together through the choices we make, the relationships we build, the communities we strengthen, and the values we embody in our daily lives. In that work of creation, every human voice matters, every human contribution has value, and every human connection makes a difference. The age of machines has begun, but the age of humanity continues.

Our task is to ensure that as artificial intelligence grows more powerful, human wisdom, compassion, and creativity grow stronger still. In that balance lies our hope for a future where technology serves not just efficiency and profit, but the full flourishing of human life in all its complexity, beauty, and meaning.

The journey continues, and we walk it together.

ABOUT THE AUTHOR

Noah Wayan is a truth-teller, heart-listener, and advocate for emotional healing and personal awakening. With a voice that blends honesty, compassion, and quiet strength, his work speaks to those who are ready to live more intentionally, love more deeply, and stop sleep-walking through life.

Known for his soulful storytelling and deeply reflective style, Noah writes for those standing at a crossroads - ready to remember who they are and realign with what really matters. When he's not writing, Noah is speaking, mentoring, and walking his own path of purposeful living - one moment at a time.

ALSO BY NOAH WAYAN

All Was Good Until You...is a powerful, eye-opening exploration of how emotional immaturity - hidden behind silence, control, and unhealed wounds - quietly erodes our closest relationships across every corner of life.

Can You Hear Me?...is a heartfelt call from men who feel invisible, ignored, or emotionally unheard. Through raw stories and powerful reflections, this book reminds us that being truly seen and understood can change everything.

Every Moment Counts...invites you to pause, reflect, and realign with what's truly important. Through soulful insights and gentle truth-telling, this book helps you step out of survival mode and into a life of intention, presence, and purpose.

I'm Ready Too!...This powerful guide shines a light on the emotional struggles men face when taught to hide their feelings. Through raw honesty and healing insights, *I'm Ready Too!* helps men break free from emotional silence, reconnect with their truth, and build deeper, healthier relationships. A must-read for any man ready to grow - and for those who love him.

The Do Book...s your wake-up call, your battle cry, your no-excuses guide to finally starting what you've been putting off. Whether it's launching that business, writing that book, going to the gym, or just making a move on a long-delayed dream - this short, sharp, and soul-stirring read will help you stop waiting and start doing.

From the Mind to the Hand...s your wake-up call, your battle cry, your no-excuses guide to finally starting what you've been putting off. Whether it's launching that business, writing that book, going to the gym, or just making a move on a long-delayed dream - this short, sharp, and soul-stirring read will help you stop waiting and start doing.

Brokeitis...(broke-eye-tis) is the #1 disease sweeping the country. It's not just about not having money - it's about the cycles, habits, and systems that keep you broke, stressed, and stuck.

In **No Help Like Self Help** : Change Your Mind, Change Your Path, Noah takes you through his own journey of chasing success while battling obstacles like comparison, fear, and the infamous “Age Game.” Through raw stories and powerful reflections, he reveals how he learned to replace poor thoughts with rich ones, give up on giving up, and step into the freedom of being his best self.